NEANDERTHAL
The Strange Saga of the Minnesota Iceman

Bernard Heuvelmans

Translated by Paul LeBlond

ANOMALIST BOOKS

*San Antonio * Charlottesville*

Neanderthal: The Strange Saga of the Minnesota Iceman
Copyright © 2016 Paul LeBlond
Afterword © 2016 Loren Coleman
ISBN: 978-1-938398-81-0
Second Edition

This is an English translation of Bernard Heuvelmans' "L'énigme de l'homme congelé" which appears in *L'homme de Néanderthal est toujours vivant*, © 2011 Les Éditions De L'Oeil Du Sphinx

Cover photomontage by Bernard Heuvelmans © Agence Martienne/ Musée De Zoologie, Lausanne

Book design by Seale Studios

For information, go to AnomalistBooks.com, or write to:
Anomalist Books, 5150 Broadway #108, San Antonio, TX 78209

CONTENTS

TRANSLATOR'S INTRODUCTION

Bernard Heuvelmans' "The Mystery of the Iceman" was originally published in French (Plon, 1974, and reprinted by L'Oeil du Sphinx, 2011) as the second part of a volume entitled *Neanderthal Man is Still Alive*. The first part was an account by Soviet historian Boris Porshnev of his "Struggle for the Troglodytes."

A troglodyte is a person living in a cave. There are troglodytes today, here and there, living in caves or rock shelters sometimes brought up to an elementary level of amenity. These are not the kind of people Porshnev was writing about. He was talking about Cave Men, our presumed primitive ancestors, whom he thought might continue to exist off the margins of human occupation, in forests, mountains, and deserts. The scientific name *Homo troglodytes* had been bestowed by Linnaeus to a second species of *Homo,* the "wild man," described by travelers. Porshnev's struggle was to convince official science and the rest of humanity of the reality of that presence and to bring us all to face the fact that we might not be the only human species on this planet.

Wild hairy men had been encountered in early Russian explorations of Central Asia; native tribes all over the wide continent knew about them. Tibetan, Chinese, and Mongolian sources mentioned them. In parallel, stories of the Yeti, encountered in the Himalayas, began to appear in the western press. For many years, Porshnev and others tried in vain to attract scientific interest in the matter and finally managed to convince the USSR Academy of Sciences to establish a Commission for the Study of the Snowman, and to obtain its blessing and support for an expedition to the Pamir Mountains, an ambitious enterprise soon diverted by competing interests.

Boris Porshnev, left; Bernard Heuvelmans, right

In Porshnev's mind, the numerous accounts of encounters with "hairy wild men" all over Asia provided sufficient proof of the survival of Neanderthal Man in today's world. In his words, "never would the study of a specimen, dead or alive, provide as much scientific information as the accumulation of eyewitness reports." Nevertheless, Heuvelmans' analysis of the frozen corpse, which he examined in Rollingstone, Minnesota, added a powerful independent confirmation of Porshnev's thesis.

The finding of the frozen corpse and its description by Heuvelmans has been heralded by some as the greatest archaeological discovery of the 20th century, the material proof of the existence of a relic population of Neanderthals in the modern world. Others have emphatically dismissed it as an outrageous hoax, an imposture unworthy of scientific attention. The controversy about the "Iceman" continues to this day.

The story begins in New Jersey, when Bernard Heuvelmans and Ivan Sanderson first hear about the frozen corpse. They quickly drive to Minnesota to have a look, but can only take photos. Attempts to obtain permission to examine the specimen more closely and to attract the interest of scientific institutions come to naught. An enquiry into the origin of the specimen triggers a bizarre adventure, involving the Mafia, the Vietnam War, and drug smuggling, giving much of the account the flavor of a detective story.

Nevertheless, Heuvelmans manages to provide a meticulous description of the Iceman, based entirely on a careful examination of the photos he took while in the presence of the specimen, and draws conclusions as to its nature based on a comparison of its anatomy with that of modern humans and fossil ancestors.

Bernard Heuvelmans was a professional scientist of some repute, although somewhat controversial. Whether the conclusions he reached in 1974, in the light of that time's scientific knowledge, are correct or not, his work continues to raise unanswered questions and, now made available in translation, may kindle a renewed interest among a broader and younger audience in the enigma of the frozen man and of the possible survival of Neanderthals.

— Paul LeBlond
Galiano Island, BC, Canada
March 2016

To the memory of my two fellow collaborators, deceased during the preparation of this work: BORIS PORSHNEV (1905-1972) who inspired much of the work presented here, and IVAN SANDERSON (1911-1973), its most passionate advocate, in the hope and with the determination to pursue our struggle until final victory is ours.

— Bernard Heuvlemans

"I have no doubt that some fact may appear fantastic and incredible to many of my readers. For example, did anyone believe in the existence of Ethiopians before seeing any? Isn't anything seen for the first time astounding? How many things are thought possible only after they have been achieved?"
— Pliny, *Natural History of Animals*, Vol. VII, 1

Chapter 1

WHAT I SAW
A creature related to both man and ape

"Even fairgrounds are areas of research for a curious zoologist. New animals have been discovered in museums and in zoos. Why not in carny shows?"
—Bernard Heuvelmans on the TV show *Sherlock at the Zoo*,
which aired on French television, March 28, 1962

To find oneself face-to-face with a prehistoric human or an ape-man is a situation normally found in science fiction stories, where time travel is possible, or in one of those adventure stories describing the discovery, in some remote corner of the world, of an area isolated for ages from the rest of the world. Nevertheless, this is exactly what happened to me, without having to use H.G Wells' *Time Machine* or to visit Conan Doyle's *Lost World*. It happened on December 17, 1968, in the middle of the United States of America.

It is such a fantastic story that I have little hope of being taken at my word, except perhaps by those who know me well, either personally or through a careful reading of my books. They know how meticulous and conscientious I am in my zoological studies; they also understand that I would not risk my scientific reputation with fantastic and thoughtless pronouncements. Fortunately, I have at hand more than just my own words to back up my story

There exist, first of all, numerous photos that have allowed me to perform a detailed study of the external anatomy of the creature, which I examined with the greatest care.

Then, there is the meaningful comparison of such anatomical data with the thousands of descriptions by people who, over the centuries, have seen or even known such creatures, similar even in the smallest details to what I have seen.

Finally, there is perfect agreement about all that is known of the anatomy, behavior, and history of this kind of creature and its distribution throughout the world, with the most reliable information from paleontology, biological evolution, ecology, and zoogeography.

Thus, there are coherent, almost monolithic, unshakable proofs in three admitted categories: *autoscopic* (which can be seen by all); *testimonial* (based on other people's reports); and *circumstantial* proofs (based on the agreement with external facts and events).

Few scientific results admitted to be on a firm basis benefit from such a robust foundation. So, you may ask, why hasn't this discovery, likely to upset current anthropological knowledge, appeared in treatises and textbooks?

It is because, quite precisely, that it does rather upset anthropology. I, Bernard Heuvelmans, should have the impudence of proclaiming and actually *being able* to demonstrate the falsehood of some scientific dogmas! (And, as I shall demonstrate, we are really speaking of dogmas, not merely opinions confirmed by some evidence.) To do so is to openly challenge this formidable institution we shall call Official Science, or also *The Scientific Establishment,* a shadowy organization always united to prevent at all cost any revelation likely to disturb the intellectual comfort of its members. For years, this Freemasonry of Science has blocked my path, as it has Professor Porshnev's, with every possible obstacle to prevent the truth from coming out. In the Middle Ages, we would have probably been burned at the stake; today, it is enough to gag us. And please do not believe that I am being melodramatic!

This book is merely the expression of the muffled cries that we have uttered through the gags, which we desperately tried to get rid of. Here is my story.

I shall present the facts in the order in which they became known to me, either because I was there or because they were related by others. This may leave a first impression of disorder and confusion. Rest assured, everything will fall into place and will satisfy the readers' curiosity. If the story seems rather tangled, it is mainly due to the fact that it was *deliberately* made to appear so in order to hide some embarrassing, even incriminating truth; also in order to hide some error or *faux-pas*; because of some financial interest; to thwart the success of a colleague; to convince oneself of one's objectivity; and finally simply by some deep-rooted love of the fantastic.

Furthermore, there were also all those who refused to acknowledge some crucial facts for fear of reprisals, had they revealed the sordid underbelly of an affair that cannot be simply reduced to a scientific enigma; and then, there are those who conveniently ignored certain aspects of the problem for fear of being ridiculed or even dishonored for having participated in this venture. People were driven by malice, pride, jealousy, ignorance, silliness, greed, and cowardice. A very human and sometimes excusable behavior, which however should not be at the cost of the truth desired by all seekers of knowledge.

I long hoped to be able to tell this story while remaining silent about some embarrassing moments, or in quickly passing over some events where good friends or famous persons played a debatable, even perhaps a shady role. Alas! I soon realized that such omissions and modesty sometimes seriously affected the clarity of the presentation and ended up transforming into an inextricable imbroglio a story that, without being very simple, is nevertheless logical and coherent.

When dealing with a problem of such great philosophical and scientific importance, it seems to me a trivial matter to hurt the feelings of a few people. After all, it's up to everyone to take responsibility for what they say and do. What matters here is to bring out the truth, as much of the truth as possible, and nothing but the truth.

In December 1968, I was the guest, in New Jersey, of long-time friend and correspondent, American writer and journalist Ivan T. Sanderson. I had arrived in New York at the beginning of October for the launch of my book *In the Wake of the Sea-Serpents* (the American version of *Le Grand Serpent de Mer: le problème zoologique et sa solution*), and I was preparing an expedition to Central America to study the local fauna and some particularly rare mammals facing extinction.

On December 9, Ivan received a call from a Mr. Terry Cullen, who introduced himself as a herpetologist, owner of a vivarium in Milwaukee, where he dealt in reptiles and amphibians. He had been extremely interested by the book and the numerous articles written by my host on the problem of the Abominable Snowman, broadly speaking, meaning the actual existence of large human-like apes still unknown to Science. This is why he wanted to tell Sanderson about the exhibition, the previous August [1967], at the Wisconsin State Fair of a creature of that kind. This creature had also been shown again more recently at the International Cattle Fair in Chicago, where one of his friends had seen it only three days before.

The creature was apparently embedded in a block of ice and looked like a rather hairy human being. According to Mr. Cullen's description, it was between 1.50 and 1.65 m [5.0-5.5 ft] tall, entirely covered by long dark brown hair, had a sagittal crest on top of its head, but no prominent canines nor opposable big toe. Overall, a description reminiscent, except for the color, of the best reconstruction of the Himalayan snowman, synthesized from the most precise details given by observers.

Mr. Cullen had also mentioned that the frozen monster had a broken skull and that its brains were oozing out at the back. The corpse was shown to the public as that of a man preserved in ice "for centuries," which suggested some great age, even perhaps, with some imagination, a prehistoric origin.[1]

According to what Mr. Cullen had been told, the enormous ice block, with its bizarre contents, has been found floating off Kamchatka, or less precisely in the Bering Sea, by a Soviet trawler. The captain of this ship had first thought that the ice contained an accidentally frozen seal, but as the ice melted, a simian shape had gradually revealed itself. Subsequently, the trawler had been forced to come into a Chinese port and the strange cargo had been confiscated by the harbor authorities, along with the rest of the shipload. The ice block then had disappeared for months before resurfacing as a contraband article in Hong Kong, where its current proprietor had acquired it. Mr. Cullen's secretary was currently working to find the latter's name and address.

The following day, the Milwaukee vivarium operator confirmed his phone

---

1   We were later to learn that the exhibitor's trailer bore this absurd sign, in the good old fairgrounds tradition:

PRESERVED IN ICE FOR CENTURIES
PERHAPS A MEDIAEVAL MAN, RESCUED FROM THE ICE AGE

conversation with a letter. He explained that because the owner, at his request, had kindly flipped over the ice block, he had been able to examine the back side of the specimen and observe some details not visible to normal visitors.

Already on December 11 Sanderson had located the man who was showing the frozen corpse, partly following another call from Mr. Cullen, partly thanks to reports from two of his Chicago correspondents who had obtained that information directly from the secretary of the Cattle Fair. He was Frank. D. Hansen, of Crestview Acres, in Rollingstone, Winona County, Minnesota.

Ivan immediately sent this person a telegram asking him to call him collect so as to arrange for an interview. However, to make sure not to alarm him by announcing a scientific inspection, he wrote in ambiguous terms, claiming that he wanted to see him for professional reasons. It turns out that many years before, Ivan had owned a small zoo and had also exhibited rare animals to the public.

Hansen called the next day to say he agreed to meet him; Ivan asked me if I would join him in the Midwest to see the specimen and provide some professional expertise. This implied a drive of more than 3,000 km [1864 miles] to get a look at what was most likely, I thought, to be some kind of trickery, either some carnival fake or a well-known animal. (The Oriental origin of the black specimen and the presence of a sagittal crest led me to think that it could be a Celebes crested macaque, a large tailless monkey also called *black ape* or *crested baboon.*) I found Ivan's confidence and enthusiasm to set out on such an expedition on the basis of rather suspect information rather excessive. However this trip would allow me to see an area of the United States that I did not know. And in any case, after twenty years of cryptozoology, I had made it a duty always to go have a look, if at all possible, at the specimens brought to my attention. I never allowed myself to be deterred by the fantasy labels—monster, sea serpent, dragon, or Abominable Snowman—with which they were usually called by laymen and especially the media. Contrary to what one might imagine, such investigations rarely turned out to be fruitless, and they often allowed me to inspect and conserve specimens of animals already known perhaps but extremely rare or unknown in some area.

This particular enquiry was to turn out to be of much greater interest, as it ended up with the description of an unknown species of hominids alive today.

Nevertheless, the affair at first presented itself rather unfavorably. It had of course everything to make it attractive to the general public: some kind of ape-man, brutally dispatched if one was to conclude from the state of his skull, but then miraculously preserved in ice from time immemorial: a picaresque story of international conflict, piracy, and smuggling beyond both the Iron and the Bamboo Curtains, the picturesque ambiance of carny shows and their monsters, and finally, an enquiry led by a pair of researchers known for their work on the Great Sea Serpent and the Abominable Snowman. There was therein every ingredient to guarantee a Hollywood

B movie success: a mixture of Tarzan, Sherlock Holmes, and James Bond, a tinge of Frankenstein, and a pinch of King Kong. Hence, there was of course enough to throw on the whole story an aura of suspicion and to make it incredible and *a priori* unacceptable.

I did not worry over much at the beginning, as I did not labor under a delusion of what we were to see, but these premises were later to become particularly embarrassing.

Ivan and I left New Jersey on Saturday December 14. We arrived Monday night in Winona, Minnesota, after a rather tedious trip. Ivan immediately called Mr. Hansen to arrange a meeting on the next day. So, on Tuesday the 17th, after wandering for hours in the remote snowbound hills of Rollingstone, we met the owner of the mysterious specimen at his ranch, or his farm, as he called it.

Frank H. Hansen was an active-looking, sturdy, forty-something, slightly tending to fat. As his name suggested, he was of Swedish ancestry. That was also noticeable through the mongoloid traits sometimes seen among Swedes with Lapp or Finnish forebears: slanted eyes that shrank to mere slits when he smiled. Hansen was a retired career soldier: a former pilot who had left the US Air Force in 1965 with the rank of Captain, after twenty years of active service. Having participated in the Korean War, and then fought in Vietnam, he told how he had spent seventeen years commuting between the United States and the Orient, which he knew well. He was married and the father of three children, of which two were also already married.

How does one go from piloting combat aircraft to the chatter of a carny? Well, Hansen had started his career as a fairground showman by exhibiting an item discovered by chance in the countryside: a relic of the first motorized tractor ever built, in 1916, the John Deere 4-cylinder. In the US, where history is so short, such items are seen as antiques! As to the object of his current exposition, Hansen told us that he had shown it in numerous fairs since May 3, 1967. He said he had bought it in Hong Kong. However, contrary to what Mr. Cullen had told Ivan, he said that the ice block had been fished up by Japanese whalers, who had then sold it legally to a Hong Kong merchant, the very one from whom he had acquired it.

The next day, after much procrastination, Hansen added that it was an agent of the Californian film industry who had first spotted the ice block during a trip to the Orient, where he was researching accessories and decorative elements. This wealthy and well-known person, whom he obstinately refused to name, was said to have given him the funds required to purchase this curiosity and to bring it back so as to exploit it for their mutual benefit. When pressed, however, Hansen conceded that the unnamed rich partner was the *actual owner of the specimen.*

On December 17, 1968, in the snowy hills of Minnesota, Sanderson and Heuvelmans paid a visit to Frank D. Hansen's trailer, where he showed them the fabulous specimen that would satisfy their wildest dreams.

The plot was thickening! Not only were there many contradictory versions of the origin of the item, a mysterious Hollywood tycoon was lurking in the shadows. But there were more surprises to come.

Hansen had frowned when Ivan had told him that he had come to see him as the science editor of *Argosy* magazine, and had become even more sullen when he introduced me as a professional zoologist eager to examine his specimen. Hansen immediately declared that he didn't want any publicity about the specimen, nor that it should be submitted to a serious scientific examination. He also explained why.

He first claimed not to have the slightest idea of the exact nature of his specimen. It was entirely possible, he said, that it might be nothing more than some clever oriental creation, like those "sirens" offered for sale in many ports of the Indian

Ocean, which are usually the product of a delicate joining of the body of a monkey or a lemur, the tail of a fish and the claws of a raptor. Actually, he admitted that, for the moment, he did not want to know any more about the said specimen, "so that he could in all honesty continue to present it to the public as a total mystery." He meant to continue to do so until came the day when, unable to preserve it correctly, he would offer it to some scientific institution, where it could then be properly examined.

This attitude didn't make any sense to Ivan and myself. Why would a man whose income was tied to the renown and success of an exhibit object to it being mentioned in a magazine that sold over two million copies? Already, a professional fairground magazine has praised it, and a number of regional newspapers had mentioned its presence at local fairs. Finally, it wasn't quite true that Hansen did not want to know any more about his specimen since he had, as he said, already submitted a few hair and a blood sample to various experts.[2]

Finally, was it really "totally honest" to present to the public the frozen creature as a man "preserved in ice for centuries" and, at the same time, as a "total mystery"?

Clearly, Hansen did not mind some discrete mention of his exhibition but was afraid of too much publicity. He also had this paradoxical tendency to depreciate his specimen: he favored the idea that it might be a fabrication, or perhaps a commonly known species, with a known type of hair and very ordinary blood.

Before granting us permission to view the specimen, Hansen demanded that Ivan give his word that he would not publish anything about it, which Ivan gave on his honor. I carefully refrained from making such a promise, knowing full well that if the creature exhibited had any scientific interest, it would be my duty as a scientist to reveal it. And, of course, why should it be forbidden to offer one's opinion on an item belonging to the public domain, seen by dozens, even hundreds of thousands of people?

All that was absurd and surely hid something else…

Anyway, apparently reassured by Ivan's formal promise, Hansen kindly allowed us to examine the specimen at our leisure. He led us to the trailer where it was kept, parked not far from his house.

In the middle of the trailer stood a massive piece of furniture, similar to a sarcophagus or a pool table, but actually more like one of those large cold containers seen in meat or milk stores, used to display perishable products. Hansen turned on the fluorescent lights mounted in the interior. And that is when we saw IT for the first time, through the glass panes that covered it.

---

2   The hair specialist was said to have answered that the hair might resemble the type of hair from an Asiatic human (?) and the blood specialist had discovered that the sample "contained red as well as white blood cells"(?). Some experts!

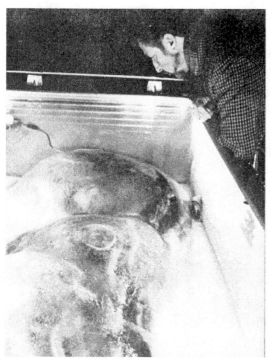

Heuvelmans spent three days studying, photographing, and sketching the specimen.

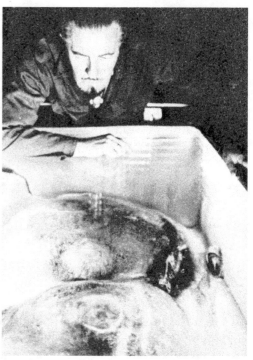

Sanderson carefully examined the specimen and took as many measurements as he could from a vertical perspective.

I would be lying if I claimed to have felt a shock, or some intense emotion, when suddenly faced with this escapee from prehistory. After all, I still had no idea of what lay under my eyes, and I was content with examining it with extreme attention, understandable mistrust, and, I must admit it, increasing amazement. What I found most astounding was that there was no question that it might be, as I had first thought, some Celebes macaque, or a siamang gibbon, or even one of the African anthropoids, also black-haired apes.

It was undoubtedly a man, tall and well muscled—a little over 1.80 m [nearly 6 ft] tall—a man who, at first glance, appeared to have the same proportions as you and I, but a man as hairy as a gorilla or a chimpanzee!

He was lying flat on his back, his head tilted back, his left arm raised over his head, his right hand seeming to protect his lower belly, in the familiar pose of a sleeping man. However, he was definitely not sleeping and undoubtedly dead. His head was stained with blood and his eye sockets were empty and bloody. His left arm was strangely curved like that of a rag doll: actually, it was certainly broken because halfway between the wrist and the elbow the ulna was sticking out of an open wound. His right foot, bent upwards while the knee was also bent, seemed in an abnormal position, resulting from a muscular spasm, a disability, or a wound. This mutilated corpse was embedded in an icy shroud.

According to Hansen, the rectangular block of ice in which it had been found originally measured 2.75 m [9 ft] in length by 1.5 m [5 ft] wide and 1.2 m [4 ft] thick. It weighed more than 6,000 pounds [2,700 kilos] when it had been hoisted aboard the ship.[3] To make the creature frozen in the ice block as visible as possible, its size had since been reduced as much as possible, and the whole front part had been cut-out and sculpted to minimize the thickness of the ice layer surrounding the corpse. In certain areas the ice matrix was very opaque, probably because of some crystallization. Odd frosty rings surrounded the right hand and the middle of the chest. In general, however, the specimen was easily visible, and numerous small details could be seen through the ice, in some places crystal clear. In some areas, however, I had to use a flashlight to aim a glancing beam of light to make out the shape of the creature.

The specimen *appeared* to be in a remarkable state of preservation. Where blood was visible, it had retained the bright hue of its liquid state, and the face still had a reasonably healthy complexion.

However, this appearance of freshness was only an illusion, as I was soon to discover. From the corner of the icy coffin closest to the left foot emanated the nauseating smell of a rotting corpse; the seals in that corner were obviously not airtight. Hansen seemed surprised when Ivan mentioned this to him the next morning. He started by claiming that this was impossible, that the specimen had always been kept at a temperature below 5°F, or -14°C (which, by the way, was not cold enough to ensure its long-term preservation!). Nevertheless, when we pointed out where the smell was coming from, he had to yield to the evidence and appeared quite upset.

A careful examination of the specimen was soon to reveal that the fifth toe of the right foot was of a suspicious grey color. When I mentioned this to Hansen, he said he had also noticed it but had not thought this to be of any importance. It was actually possible that the toe had already become grayish *before* the death of the creature. I already mentioned that the position of the right foot seemed to suggest some disability or even a wound. The grayish appearance of the little toe might have been caused by a gangrenous festering wound.

Hansen thought he could continue to show his hairy man for another year, but there was a distinct possibility that over that time most of the fleshy parts of the specimen would putrefy to such a degree that an in-depth scientific study of its histology, serology, and immunology would become impossible. In other words, it would become impossible to study the nature of the tissues, to determine the blood-type, to perform an analysis of the blood proteins by electrophoresis, and, most importantly, to perform a blood precipitation experiment with specific antibodies, an essential technique to determine the specimen's relationships with other creatures.

---

3   Ever suspicious, I calculated the weight of a block of ice of those exact dimensions at 4,500 kilos [9,920 lbs]. Either the ice block was more irregularly shaped or, perhaps, it had never existed.

That seemed to me rather a pity, since the more I examined and studied the specimen, which I did for eleven hours over three consecutive days, the more authentic and valuable it appeared to me.

As a precaution, on the second day, I took many photos of the corpse, in color as well as in black-and-white, which wasn't always easy and required some contortions. The light shed by the lateral neon tubes within the sarcophagus was very dim and had to be enhanced with a powerful lamp backed by a reflector, which Hansen was kind enough to lend me. Because of reflections off the ice, I had to use a polarizing filter, which I had bought on spec in Winona. In addition, the low ceiling of the trailer made it impossible to obtain a view of the entire body from above; using a wide-angle lens would have created unwanted distortions. Armed with my Asahi Pentax, loaded with Ektachrome High Speed, I had to be content with oblique or partial vertical views. Using four of the latter, end to end, I finally managed to create a composite of the entire specimen (see cover).

A revealing incident occurred during my photography session. At one point, while the high-power lamp was just above the glass cover of the coffin, there was a sharp cracking sound: the heat from the lamp had cracked the glass.

"Good Lord!" shouted Hansen. "How will I be able to explain that?"

Hansen suddenly looked annoyed and worried, and his reaction had been too rapid and spontaneous to be feigned. Even before admitting to us that he was not the real owner of the specimen, I knew that he had to account to someone else for it.

But for now, let's return to the specimen.

What could that mysterious creature be, surrounded as it was by so many mysteries? I said that at first sight it looked like a man of normal stature and proportions but extremely hirsute. But, in fact, he was not merely an individual suffering from an abnormal amount of hair-growth—hypertrichosis, as it's called. One quickly noticed that he was rather extraordinary in many other ways, as a detailed description will show.

Except for his face, the palm of his hands, the sole of his feet, and his genitals, the creature is completely covered with dark-brown hair, generally from 7-10 cm [2.7- 4 in] long, but reaching 15 cm [6 in] in some areas, especially on the back, if one is to judge from the hair that sticks out on his sides. The skin has the waxy and ivory tint typical of untanned white-skinned people. The epidermis was easily seen on the whole visible surface of the body, particularly in the naked areas in the middle of the chest, in the armpits and groin, and on the knees, individual hair being often as far as 2-3 mm [0.08- 0.12 in] apart. Overall, the pelt was reminiscent of that of an anthropoid ape, not like the thick fur of a bear, for example.

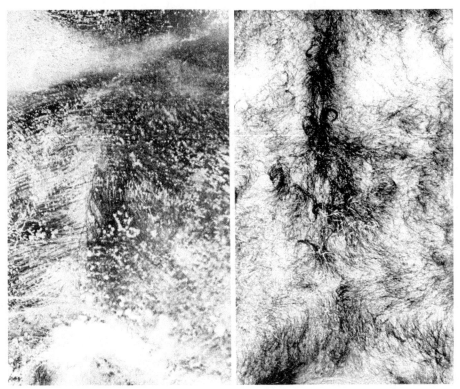

Although certainly a hominid, the specimen was as hairy as a chimpanzee, but it was not at all like an excessively hairy human, as one can see in a comparison of the specimen's chest (left) with a hairy human chest (right).

The face is essentially hairless. There is some hair in the nostrils, thin eyebrows, and a fringe of eyelashes along the eyelids. There are a few short scattered hairs on the cheeks, arranged a bit like a cat's whiskers. What is striking in such a hairless face is the ridge of short hair running along the septum of the nose, from its base to its upper end.

There is neither mustache nor beard; some longer hair appear along the weak chin and on the sides of the lower jaw, making a dark collar, but they are more like the beginning of the hair on the neck than a real beard.

Given the backward tilt of the head, there is little to be seen of the fleece that seems to cover it beyond the eyebrows. It is not possible to say whether it is made of hair similar to those of the rest of the body, or of longer head hair.

Two points of particular interest: first, the chest is partly bare, as in anthropoid apes. The hair is bent sideways on either side of the sternum, leaving a central stripe where the skin is more visible than elsewhere. Second, the tops of the feet are also covered with thick hair, a trait not observed among anthropoid apes, which have nearly naked feet.

The anatomy of the creature is even more bizarre than its hairiness.

In spite of the thrown-back position of the head, which would in a human being make the Adam's apple most prominent, the neck, deep in the ice, is hardly visible. Perhaps this is due to the fact that this creature carries its head deep into its shoulders. The throat is swollen, perhaps due to a goiter, a pathological condition, or perhaps because of the presence of a vocal sac, as one finds in the orangutang or the siamang gibbon.

The chest is much more bulging than in *Homo sapiens*; the collar bones are quite curved, accentuating the swelling of the chest. The chest blends into the abdomen to give the trunk a barrel shape somewhat similar to that of the anthropoid apes. The rosy nipples are found at the same place as they are in modern man.

The penis is thin and pointed and resembles more that of a chimpanzee than that of a man. The scrotum is rather small.

The arms, once measured, turn out to be rather long. They are about 90 cm [35 in] long and must reach down to the knees when loose along the body. One should note, however, that this is due to the inordinate length of the hands rather than that of the arms, and especially the forearms, which are rather short.

At first sight, the legs also seem to be quite long, but that is merely an illusion. Usually, when one sees large hairy primates, they are anthropoid apes, which have *much* shorter legs. Actually, our hairy man has rather short legs (90 cm from the waist down), but that length remains within the normal range of Modern Man.

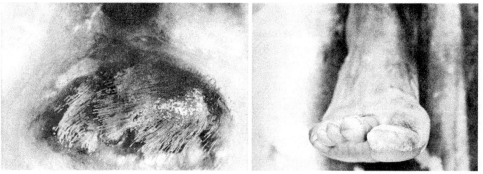

The specimen's foot was extraordinarily wide with crooked toes of nearly uniform thickness (left); it did not resemble a human foot (right).

The most striking features of this creature are its hands and feet: elegant and refined hands and crooked toes. These appendages are as large and almost as massive as those of an adult gorilla! For the sake of comparison (I am 1.73 m [5 ft 10 in] tall, and thus slightly above average at 1.65 m [5 ft 6 in] for our species), my hand is 19 cm [7.5 in] long and 8 cm [3.1 in] wide, but the hairy man's hand is 26 cm [10.2 in] long and 12 cm [4.7 in] wide. The length of the creature's foot cannot be accurately determined, but a calculation estimates it at 25-27 cm [~10 in], a normal length for a 1.80 m [5 ft 11 in] man. Its width is, however, quite disproportionate! My foot has a maximum width of 10 cm [4 in]; the creature's foot is more than 50% wider at 16 cm [6 in].

The hands and feet are not only remarkable by their size but also by other original features.

Let's look first at the hands. The thumb is unusually long, for men as for apes. In anthropoid apes, the thumb is much shorter than in modern man, but in contrast, the hairy man's thumb is longer. Its end, when side by side with the index finger, reaches the level of the joint between the last two phalanges.

Furthermore, as one may note from the position of the fingernail, the thumb does not appear to be as completely opposable as that of our species. It is as if during embryonic development the gradual rotation of the thumb had stopped at a stage somewhat earlier than that achieved in *Homo sapiens.*

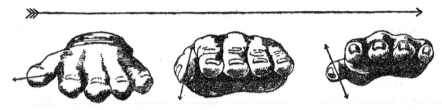

Progressive twisting of the thumb through the development of the human embryo (after Schultz, 1931).

The reduced opposability of the thumb is also reflected in its greater length, which gives it a thin, cylindrical shape. In our species, the thumb tends to flatten towards the end into a spatula shape, which allows it to cover a wider area when opposed to the other fingers. In this way, we end up with a strong and effective grasping precision tool. There is nothing like this in the hairy man; his thumb looks just like the other fingers, long and narrow. His hand looks more like a scythe than pincers. He basically is endowed with lower extremities adapted to gathering grasses or leafy branches or to pull out roots and bulbs.

We should add, however, that the fingernails are real nails, and not claws. They are relatively narrow and rounded across, like old Roman tiles. The fingernails are thick and yellowish. It is striking to see in the frozen specimen how the nails stick out beyond the fleshy end of the fingers by a few millimeters and by as much as a centimeter [0.4 in] in some cases.

Let's now look at the lower limbs.

Endowed with a non-opposable big toe, the hairy man's foot is typically human—that anatomical detail is actually considered the safest way to superficially distinguish Man from Monkeys and Apes. The foot is, however, not at all like that of Modern Man.

As we have already noted, it is relatively short and stubby[4] although large in absolute dimension. Besides that, the toes are much more uniform in thickness than ours. The "big" toe, sticking out somewhat from the others, although still non-opposable, is not much larger than the other ones. The weight of the body does not rest mostly, as it does with us, on the inner side of the foot. The toes stick out straight along the axis of the foot rather than slightly obliquely, as ours do. The toes fan out; the second and the third (rather than the first and second in *Homo sapiens*) are the longest.

All the toes are strangely bent down, and the smallest toe seems folded towards the inside. This kind of grabbing structure should provide an excellent grip on the ground, particularly in rough terrain. It is not the foot of a savannah runner, like that of *Homo sapiens*, nor that of a tree-climber as that of apes, but rather the foot of a rock climber, a mountaineer.

The sole of the foot is much more wrinkled and divided into pads than that of Modern Man and, in that regard, is closer to that of the great apes. Whether it be to hang on to branches or to rocks, such a structure provides well-known anti-slip properties, as one can see on automobile tires.

The toenails are, like the fingernails, also rather narrow and yellow. They seem rather thick and curved across, which would increase their strength. Their structure contributes to a better grasp on rough ground; they are the points of that grappling hook of a foot.

Finally, let's look at the head.

What is most striking about the face is how enormous it is, in absolute terms, as well as how disproportionate it is compared to what we can see of the rest of the skull, namely the barely visible receding forehead. Overall, the head is no higher than that of *Homo sapiens*, but it is nearly fully occupied by the face, which is surprisingly extensive. The forehead is low and receding, which is why it is barely visible. The sloping forehead makes the eyebrow ridges stand out.[5] They seem to form a continuous swelling over the eyes.

Unfortunately, the eyeballs are not in their orbits. (One can, however, make out the shape of one of them on the left cheek.)

The cheekbones appear quite prominent, at least from the side, which implies a strong development of the zygomatic bones.

The nose is undoubtedly the most characteristic feature of the hairy man's face. In shape, it is more reminiscent of that of the *Rhinopitecus* (the snub-nosed monkey)

---

4   Among the Alakaluf of Tierra del Fuego, who go barefoot in snow and icy water, there is a tendency for the foot to take on such proportions.

5   If there was a sagittal crest on the top of the head, as claimed by the original informer, it might be visible. Mr. Cullen's description was, of course, rather vague, and even incorrect, especially with respect to the height of the specimen, which he underestimated by 15-30 cm (6-12 in).

than of that of any modern human race. Not only is it extremely wide, but it also highly upturned. The tip of the nose reaches nearly up to the somewhat sunken root of the nose, at eye level, so much so that the ridge of the nose is nearly horizontal. The lower surface of the nasal organ is abnormally developed, featuring two gaping circular nostrils opening forward. It is the nose of a creature for whom the sense of smell must play an important role.

This incredible and rather comical looking nasal apparatus is flanked on either side by the labio-nasal folds. Further out from the nose, the parallel oculo-malar folds are less pronounced but stretch as far down as they do below the mouth. There is a lot of room below the nose but—and this is an important detail—there is no trace of the philtrum, the naso-labial cleft that characterizes all modern humans.

The mouth is straight and wide but has no lips, another feature which differentiates this hairy man from all modern human races. It is slightly ajar and one can discern a yellowish tooth that seems to be the upper right canine; it doesn't seem particularly long or pointed. In short, it does not stick out beyond the row of the other teeth, as it does in monkeys.

Seen from the front, it is difficult to assess the degree of prognathism of the face, but it appears low. The face is essentially flat.

The lower jaw is rounded and sloping; there is no real chin, or "chin projection" as anthropologists would call it.

The ears seem pointed but, as one can only make out their outline, it is difficult to tell whether it is the lobe itself that is pointed or whether a tuft of hair makes it look that way.

Oblique view of the specimen's head, showing the shape of the nose (artist interpretation from a photo).

In summary, the creature lying in front of me within its lit-up freezer sarcophagus presented a curious mixture of human and simian characteristics.

With its rich pelt and the arrangement of its hair, it resembled the great anthropoid apes, particularly the gorilla and the chimpanzee. The weak opposability of its thumb made it closer to some lower primates, such as the New World monkeys, in particular the capuchin monkeys. Its lack of lips and philtrum distinguished it, in the company of all other living primates, from our own species. However, the proportions of its limbs was very similar to that of many modern men, and also, one should not forget, to some New World monkeys. Its well-developed nose was clearly human-like, but its shape was reminiscent of that of a snub-nosed monkey, the *Rhinopitecus*. Its voice pouch, if he indeed had one, would make it nearer some Oriental apes, in particular the orangutan and the siamang gibbon.

That said, the presence of a non-opposable big toe, weakly developed canines, and a general anatomy typical of a bipedal vertical stance, undoubtedly placed it in the ranks of hominids. All things considered, it was simian only in some external features, skin-deep or related to near-surface muscles.

So, as one speaks of a real monkey showing a superficial resemblance to a man as "anthropoid," one could call this creature "pithecoid," meaning a real man, with only a superficial resemblance to a monkey.

Chapter 2

WHAT IT WAS NOT

Neither a well-preserved fossil, nor a drifting Ainu, nor a hairy monster,
not the hybrid of an ape and a woman, nor a simple fabrication

"At most, what we could hope to expect would be that during the last ice age,
some prehistoric man or other would by accident have fallen into a bog
in a permafrost area, as in Siberia, so as to reappear some day
frozen and still fresh, like one of those mammoths."
— William Howells, professor of anthropology at Harvard University
and president of the American Anthropological Association,
in his book *Mankind So Far*, 1944

Many questions come to mind when one comes face to face with a creature that is partly man and partly ape, as was the case for the specimen in the hands of Frank Hansen, in a remote corner of Minnesota.

Was it a prehistoric man, or perhaps an ape-man preserved in ice, as claimed his promoter, not merely for centuries but more realistically, given the circumstances, for millennia. Might it not be a "monster," meaning an abnormal specimen of our own species? Or perhaps the fruit of some bizarre affair between an ape and a woman, or a man and a she-ape? Or might it simply be an Ainu, commonly called "the hairy Ainu," a representative of a well known but declining human race? Or was it merely a hoax, a manufactured dummy or a composite creature made up of carefully joined parts from limbs and organs of living creatures of different species, or perhaps a mixture of natural and fabricated parts?

All these questions came to mind, or were asked of me, and I have mulled over them for weeks, months, and by now years, and I cannot come up with a definitive answer.

First of all, one thing is certain: it was physically impossible for this specimen to have been preserved in ice for very long. It is a fact that the rate of decomposition of a corpse depends on the ambient temperature. But, while it is true that one may artificially lower the temperature to a point where putrefaction is completely inhibited, such intense cold is not to be found on our planet, at least on a permanent basis.

Ah! But, one might object, how about those Siberian mammoths? The fact is that they were not discovered in ice, as is usually reported, but in a frozen muck replete with rotten vegetation. Because of the presence of bactericide acids, like tannic or humic acid, this kind of peat has a tendency to prevent decomposition. The cold of the ice can only slow down, not completely prevent the process.

If instead of having been recently frozen, the hairy man specimen had been preserved for thousands of years in peat, bitumen, or some such substance, as is the case for some fossils, it would have acquired a characteristic stain, tanned, desiccated, and mummified. (I am now thinking of the famous Tollund man, found nearly intact after spending two thousand years in a peat bog of central Jutland in Denmark.)

That is not the case here, and one can firmly state that the hairy man has been frozen for at most a few years.

The particular structure of the ice, with a countless string of bubbles caught all over the surface of the frozen corpse, leads us to conclude without doubt that *the creature was frozen artificially*, and not spontaneously following an accident, for example. In nature, the freezing of such a being in water would have begun at the skin, and the layer of ice formed there would have tended to gradually expel outward any gases dissolved in the surrounding liquid.

In a freezer, exactly the opposite takes place. Ice begins to form at the surface of the freezing elements, which are usually placed on the sides of the freezer, and at any air-water interface. The gases dissolved in the freezing water escape into the remaining liquid caught within the icy box, and they end up making bubbles within it. Have a look at an ice block, or at the ice cube that you drop into your whisky: they are clear and transparent on the outside, but opaque in the middle. Of course, should a foreign body be immersed in a freezer full of water, the bubbles would then accumulate all over the surface of that body. And this is exactly what one observes here.

Anyway, a quick glance would allow recognition of natural ice. Natural ice, created as it is within a moving liquid, is subject to continuous mixing, which traps dissolved gases trying to escape, and thus is almost always opaque. When it is not, as for example in icicles, it is because it has been formed very slowly, drop by drop, giving dissolved gases the opportunity to escape completely.

In summary, the corpse had certainly not been found as such in a block of ice drifting at the surface of the Bering Sea, or anywhere else for that matter. It had been carefully set on ice in a freezer, just like a piece of meat.

The hairy man was then neither prehistoric, nor even mediaeval. Actually, judging from his wounds, it is most likely that he had been shot. According to Hansen, who had been able to see the dorsal part of the specimen before it was placed in its current container, the whole back of the skull was bashed in or missing and the brains were oozing out.

Besides, the right eye socket was empty and bloody, and the left eyeball was out of its socket (its shape was just visible next to the cheekbone). From that evidence, a forensic expert would most likely have concluded that the victim had been shot in the face by a heavy gauge weapon, or perhaps point blank by a lighter weapon.

In any case, the bullet had gone through and destroyed the right eye, and the shock wave had blown the other eye out of its socket and left a large crater at the back of the skull; death must have been instantaneous.

One might even be tempted to re-enact the crime.

As for me, the sum of all the observed clues led me to think that the hairy man had been caught alive after being immobilized by a shot to the right leg. The bullet would probably have hit one of the sciatic nerves controlling the muscles that stretch and flex the foot. That would be the only way to explain the abnormal bending of the right foot in the context of a general muscular relaxation of the rest of the body. The victim probably ended up lame after its capture. Gangrene might even have begun to infect the wounded leg.

His pain must have made him irritable and, given its muscular strength, quite dangerous. He must have remained captive for quite a while, given the abnormal pallor of his skin and the length of his fingernails. Perhaps he tried to escape or attacked its jailers. He might then have been struck with a stick or a metal bar, and his left forearm would have been hurt while trying to fend off the blows. Panicked at the sight of this wild, enraged creature, someone had in the end shot it. It was also possible that the left arm had first been broken by a bullet.

One might suppose that its captors intended to sell it to a zoo or more likely, given its human appearance, to show it as a freak in fairgrounds. Once dead, they had decided, so as not to lose all profit from its capture or purchase, to freeze it and to cleverly present it to the public as a man frozen in ice since time immemorial.[6]

All that is speculative, of course, and could be verified only through an autopsy, but this explanation would account for all the bizarre features of the corpse without imagining unrealistic hypotheses.

So there could be no doubt that this creature was of the here and now. But was it a member of one of the known races of humanity?

The only modern men who are reputed to be extremely hairy are the Ainus, a nearly extinct white race whose last hundred-or-so members live on Hokkaido Island. Not so long ago, they were also found on Sakhalin Island as well as in the Kuriles. Since the specimen in question had been discovered off Kamchatka or near the Bering Strait—at least according to the story—one might be led to believe that it could be one of those Ainus, whose frozen corpse had been rafted north by an ocean current. During my conversations with Hansen, he wondered about this possibility. Which goes to show that a layman might well have considered this hypothesis as likely. But it was also possible that the supposed origin of the specimen might have been in the Bering Sea itself because of its proximity to the islands where the last of the Ainus live.

---

6   Perhaps had they read Prof. William Howells' book from which I cited a relevant passage at the beginning of this chapter.

Unfortunately for this explanation, or scenario, the oceanic current flowing along the Kurile Islands and the Kamchatka Peninsula, the Oyashio, flows southwards from the Bering Sea towards Japan, and not the other way. Anyway, the Ainus' reputation for being hirsute is much overblown. They may be very hairy compared to other Orientals but not with respect to other human races. As Dr. Carlton Coon points out in *The Origin of Races* (1962), the Ainus "have no more nor less hair than a hairy Scot or Jew." Nothing like the extravagant hirsuteness of Hansen's specimen.

If the specimen is not a normal representative of a known race of modern humans, might it not be an abnormal sample of one of these races, a "monster" as one might say? Many examples of extreme hairiness, hypertrichosis, have been recorded over the centuries in the annals of medical literature. Some of them, like Lionel, the Man-Dog; Julia Pastrana, the Spanish bearded lady; and Kra, the Vietnamese girl, have had their moment of fame. Most of these curiosities, especially the bearded ladies, make a living on fairgrounds. It would be quite normal to imagine the hairy specimen to be one of those.

However, if one takes the trouble of reading from cover to cover the classic treatise by Drs. Le Double and Houssay on hairy people (*Les Velus*, 1912), one finds nothing, at least in the details, like what is seen in the present specimen. Two crucial facts must be pointed out.

First of all, in the various known kinds of human hypertrichosis, hairiness is particularly abundant i*n those areas where it tends to develop in normal individuals*, namely first of all on top of the head, the chin and cheeks, the upper lips, the armpits, the middle of the chest, and the pubic area, and to a lesser degree on the forearms, the back, and the buttocks. However, in this specimen, the middle of the chest and the armpits, together with the knees, are the least hairy areas of the body (excluding of course the face, the palm of the hands, and the soles of the feet). Besides, the pubic area does not seem particularly hairy since the scrotum is nearly bare. As to the face, it is nearly beardless.

Secondly, the tops of the feet are as hairy as the legs, which to my knowledge has never been observed to such a degree in any case of hypertrichosis.

Might it then be some completely new form of hairy aberration? One might think so, if it weren't for the fact that the hairy man exhibits *so many* other anatomical anomalies with respect to modern man. Think only of how enormous the hands and feet are! Some anthropologists have suggested that this specimen might exhibit a case of acromegaly (a degenerative disease caused by excessive production of growth hormone secreted by the lower lobe of the hypophysis, and which results in gigantism and excessive development of feet and hands). However, this hypothesis fails on the fact that acromegaly is also characterized by a pronounced thickening of the lips and excessive development of the nose and chin. The nutcracker profile of Polichinelle [a French puppet figure] is typical of this condition. All this is in complete contrast with what is seen on the Hansen specimen: a short nose, absent lips, no chin.

Anyway, no one has ever seen, as far as I know, a case of combined acromegaly and hypertrichosis. The situation would be even further complicated, in this creature with nary a forehead, by a case of microcephaly. Such an accumulation of "monstrous" features is quite unlikely in the same individual, a monster among monsters!

That being said, couldn't such a monster, a veritable blend of monstrosities and a kind of living catalog of various aberrations, be the result of a deep and fundamental disturbance of its genetic material? Isn't this the chromosomal chaos that one would expect from unnatural unions between different species? It wasn't entirely illogical to think that an ape-man, a creature at once ape and human, could be the result of the union of a man and a female ape, or an ape and a woman...

I have to confess that, at the time, this idea never even occurred to me, most likely because I saw it merely as a fantasy in the style of *The Beauty and the Beast*. It is commonly assumed in modern anthropological circles, likely as a reaction against the outmoded ideas of past centuries, that hybridization between man and beast is absolutely impossible. However, not only is there no shadow of a proof that this is indeed the case, but there are signs that suggest that it might be possible. Anyway, the influence of accepted science, a network of beliefs so hardened and petrified that they are taken as postulates, is so pervasive that even I, who has stood as the champion of a science without blinders, open to all possibilities, occasionally yields to preconceived ideas. So, when I was writing my preliminary scientific report on the frozen man,[7] and reviewing all that he might be, I forgot to suggest that he might be a hybrid. An absurd hypothesis, completely unlikely, unworthy of consideration, in the eyes of most of my scientific colleagues. Nevertheless, to ignore this possibility was a scientific error, an unforgivable omission, as I was to be reminded by a letter I received soon afterwards.

Following publication of my report, a lady who wished to remain anonymous told me this strange tale:

"Around 1952-53, a reliable person told me about meeting at some friends' house a physician recently escaped from a Siberian gulag; this man was spending a few days in France while waiting for a visa to the USA. He had been arrested for insubordination: he had refused to perform artificial insemination of Mongol women with the sperm of a gorilla.

"The Russians have created a race of ape-humans: they are on the average 1.80 m [5 ft 10 in] tall, hairy over their whole body and are very similar to the specimen you describe. They work in the salt mines; they possess a herculean strength and work

---

7   Heuvelmans, Bernard, 1969. "Note préliminaire sur un spécimen conservé dans la glace, d'une forme encore inconnue d'Hominidé vivant: *Homo pongoides* (sp. seu subsp. nov.)" *Bull. Inst. Roy. Sci. Nat. Belg.*, Brussels, 45, no. 4, 10 Feb. 1969.

without pause; they mature faster than humans and are sooner ready for work. The only problem is that they do not reproduce, but the researchers are working on that.

"All this is a closely guarded secret; they never go out, and to avoid any problem, they are taken away from their mother at birth."

In spite of the rather vague nature of the information, received third hand, and its aura of anti-Soviet propaganda, it would have been incompatible with the spirit of Science to dismiss without further ado the proposed explanation. Furthermore, the story included some minor details that rang true to the ears of a zoologist.

First of all, one should point out that a crossing of anthropoid apes and humans is not theoretically impossible from the perspective of genetics, especially if one speaks of the two great African apes, the gorilla and the chimpanzee. Hybridization requires a sufficient degree of similarity between the genomes of the potential parents, namely in the number and structure of their chromosomes, as this is actually the case. One can readily imagine that a human being with 46 chromosomes might, when coupling with an anthropoid ape with its 48 chromosomes, yield a creature with 47 chromosomes, which would be sterile since it would have an odd number of chromosomes (like the mule, which has 63 chromosomes, the offspring of the horse which has 64 and the donkey with 62).

On the other hand, it is well known to agronomists and husbandry specialists that hybrids of different species or even genera exhibit extraordinary vigor, referred to as "hybrid vigor" or heterosis. Furthermore, with the rate of maturation in humans being particularly slow compared to that of anthropoid apes, it is logical to expect that an intermediate creature would grow to adult size faster than a human would and consequently would also attain more rapidly the level of muscular development required to perform their work.

So one would expect that a bastard of Man and Ape would be sterile, and would manifest extraordinary vigor and accelerated growth, which is precisely the flaw as well as the advantages attributed to the "ape-men" of Soviet salt mines mentioned by my correspondent.

In addition, one should recall that such attempts at hybridization were actually carried out, for example in French East Africa in 1926-27 by a team of Soviet researchers led by Dr. Ilya Ivanovitch Ivanov (1870-1932). As far as one knows, given the smoke screen that masks all such research on "moral and religious grounds," the experiments were not successful.

Nevertheless, the hybrid hypothesis must be rejected in this case. Why? It's not easy to give a simple answer to this question. One must admit that of all the possible but discarded explanations, it is the one most difficult to refute categorically. It can be discarded only *a posteriori*, in the light of what is generally known about wild hairy men.

When hybridization is possible between two different species or genera, the offspring show a mixture of traits from both of their parents. However in each of these offspring, the relative dosage of the mixture varies: some take more after their father, others after their mother, and we can imagine all in-between variations. In short, the offspring are different from each other: a population of hybrids is, by definition, heterogeneous. Which means that faced with a single individual that cannot be fitted within a currently known category, it is impossible to be sure whether one is dealing with a representative of an unknown type, or the result of an interspecific or intergeneric coupling. One can be sure that the specimen is not a hybrid only if one knows a sufficiently large number of very similar individuals constituting a homogeneous population. In this case, that would have been a case of putting the cart before the horse. The matter is to prove the existence of an unknown species by studying one of its members, while the representativeness of that individual could be ascertained only if the existence of the species had already been established. The only way out was to accept, as a working hypothesis, that the specimen in question was not a Man-Ape hybrid, which was confirmed in the end.

That said, it was imperative to discover as soon as possible whether Hansen's hairy man might not simply be a fake, which was obviously the explanation that came to everyone's mind.

To ascertain that the specimen was not the product of some faking was unfortunately not possible at the time of our initial examination: there was no question of obtaining a biopsy. However the likelihood of a hoax seemed so improbable to me that, even before I had been able to amass an array of irrefutable proofs of its authenticity, I felt quite safely that I could discard that possibility. I had spent days and nights reviewing that option, trying to imagine how one could artificially reproduce what I had seen. *In theory*, it was of course possible—one can imitate anything—but *in practice*, I could not see how one could reach such perfection in the art of forgery.

Just think about it! To manufacture from scratch such a fake would have first required the construction of a mold, made out of rubber, or wax, or some synthetic material, carefully colored so as to reproduce details as fine as the papillae and skin pores, together with wrinkles and folds corresponding to muscles and joints, veins just below the skin and all the small defects likely to be found: scratches, scars, moles, and pigmented spots... About half a million hair would then have to have been implanted at the correct angles (the observed hair trends being strictly corresponding to those seen in humans and anthropoid apes) and all those subtle disorders such as wounds, abrasions, blood stains would also have had to be imitated. If that was really what had been achieved, the result was more perfect than anything ever done by the best specialists, be it at Madame Tussaud's wax museum in London or the Musée Grévin in Paris, for the greatest natural history museums or for the best fantasy films. Furthermore, to improve on this masterpiece to the extent of faking a smell, one would have had to include some meat, which would have gradually rotted, and to make sure that the gases generated by the putrefaction would not have blown up the specimen.

One might imagine that it would have been easier for a forger to make up a composite being from parts of various animals and let nature take care of the rest. Let's not delude ourselves. Just to make sure that the stitches remain invisible would probably have been child's play; oriental counterfeiters are devilishly clever at such a skill. Finding all those diverse parts would actually have met insurmountable difficulties.

In order to create such a specimen through a *post mortem* surgical operation, one would have to gather anatomical pieces from at least three different creatures: the head of a man, the torso and limbs of a chimpanzee (because of the pallor of the skin and the color of the hair of that anthropoid ape), and the hands and feet of a gorilla (because of their size). However the result of such a collage would still be far from looking like our specimen.

Assuming that it were possible to find a microcephalic man without lips and without a philtrum, and with a very unusual upturned nose, it would have also been necessary to implant a few hair on the septum of his nose and in the middle of his cheeks. That would have been the simplest part of the task, but one not to be forgotten.

The problem with chimpanzees is that their skin is white only when they are young, and thus small. Their skin darkens with age, uniformly or in spots, and when they are older, it is *always* more or less pigmented. And where would one find a chimp with legs as long as a man's? Actually, one would have to graft to a chimp's torso the legs of a hypertrichosic man with equally long hair of the same color.

Finally, the hands and feet of the gorilla would have to be significantly modified, mainly because that ape's skin is black. The only known gorilla with white skin is *Copito de Nieve* (Snowflake), the albino gorilla [1964-2003] of the Barcelona Zoo. One would have to find another one like it, an adult, and then dye its hair black. And that's not all! One would also have to cut off its thumb—too short—and replace it by a longer finger: a chimpanzee's middle finger, for example. As for the feet, one would have to bring the big toe closer to the others so as to make it non-opposable, and to stretch it a little to line it up with the other toes.

So, all that would have to be done would be to bring together an abnormal man with a very bizarre face, another with severe hypertrichosis, a chimpanzee having kept its skin exceptionally white into adulthood, and an albino gorilla—three or four monsters—to fabricate through a number of grafts a creature similar to Hansen's hairy man. In other words, a most exquisite corpse in the best surrealist tradition!

A much simpler method of fabricating such a specimen was nevertheless discovered by Ivan Sanderson, who was a very experienced taxidermist. As a young zoologist, he used to skin, prepare, and mount for the British Museum of Natural History the specimens that he collected during his expeditions.

While expressing his firm conviction that the specimen that we examined was not the product of a falsification, Ivan claimed that he could nevertheless construct an exact replica. All that he required was the fresh corpse of a large pale-skinned chimpanzee. It was then possible to stretch that body so as to give it more human proportions. The skin of the head would then have to be fitted over a (microcephalic?) human skull. As to the skin of the hands and the feet, it could be stretched and deformed at will with the help of a glove-spreader, an instrument that would also be used to give the fingers and toes the required length.

All this was perfectly possible, but Ivan seemed to have forgotten one point: as I pointed out earlier, *there are no clear-skinned large chimpanzees!*

In order to accomplish the taxidermic prowess that Ivan proposed, it would have been absolutely necessary to find an albino adult chimpanzee—or even better, an adult albino gorilla whose skin would not need to be as distended—and then to dye its fur black. However, if one were to possess such a rare and valuable specimen, it would be worth exhibiting for its own sake. It would be quite stupid to use it as the basis of a suspiciously fake exhibit!

A variety of psychological arguments also spoke against the fake body hypothesis.

First, why would a counterfeiter bother simulating wounds? These wounds—empty, bloody eye-sockets, a sore in the middle of the eyebrows, an open fracture on the left arm—created the kind of horrifying atmosphere loved by the crowd, but they eventually made the specimen less impressive. As far as being spectacular, the absence of eyeballs was particularly lacking: imagine the impact created by the frozen man staring out of his icy coffin! Of all prosthetic organs, eyeballs are the most easily and perfectly imitated, and easily found commercially. If that was a fabricated specimen, why deprive it of such a dramatic element?

There is an even more serious objection to the counterfeiting hypothesis.

If a counterfeiter were to construct such a specimen, it would obviously be either to stun onlookers for profit or to maliciously mystify the experts. For that purpose, the hoaxer would have most likely created a "prehistoric man" or an "ape-man" in line with scientific, artistic, or popular ideas of such creatures. He might even have tried to emphasize those characteristics that distinguish it from modern man. But there is here no such thing. The most simian traits attributed to humans and subhumans (low forehead, prognathism, bulging eyebrows, crooked legs, etc.) are not to be seen or hardly are noticeable here.

Had the hairy man conformed more closely to general ideas of the appearance of a prehistoric or ape-man, it would certainly have had much more success on the fairgrounds than it actually had. The press would have seized on it, as it did with Barnum's most famous attractions, and specialists—anthropologists, zoologists, palaeontologists, physicians, etc.—might have rushed to examine it. The reason that for a year and a half it had not attracted more attention than the usual fairground exhibit, hardly more than an amusing side show, was that it was not impressive enough, and that it did not correspond to the traditional image that a counterfeiter would have strived to reproduce.

It's just a matter of basic common sense: if one wants to market a fake Vermeer painting, one draws inspiration from Vermeer, not from Watteau or Greco! What counterfeiter would imagine creating a fake that did not resemble any model? That would be absurd. Could one even speak in that case of a fake? A fake what?

Thus, if Hansen's specimen was the result of some fabrication, how could one reconcile the meagerness of its inspiration with the striking virtuosity of its execution? It was hard to imagine a counterfeiter at once stupid and brilliant, a flagrantly clumsy virtuoso.

I found myself once more mulling over one of my favorite aphorisms from my master, Sherlock Holmes, which is also my favorite quote: "When you have eliminated the impossible, whatever remains, however improbable, must be the truth."

Of course, but just how many hypotheses could there be? How many could one *a priori* imagine to explain the existence and the characteristics of the specimen exhibited by Hansen?

For me, six hypotheses, no more no less, covering the whole range of possibilities, deserved consideration *even before a detailed examination of the specimen*. The specimen could be:

> 1. A fake, that is either an object entirely manufactured and artificial, or a combination of parts assembled from various animals, or a combination of the two;
> 2. An ordinary human being, meaning a normal person belonging to one of the known races of modern Homo sapiens, but to a poorly known one since it could not be identified;
> 3. A monster, meaning an abnormal individual, the result of some genetic or embryological accident, a teratological specimen;
> 4. A hybrid: a normal individual resulting from the coupling of individuals belonging to different species or even genera;
> 5. A fossil man, but not really fossilized, meaning an individual belonging to some extinct form of mankind, preserved in flesh and bone for millennia due to its burial under exceptional circumstances;
> 6. An unknown man, a normal individual belonging either to a still unknown race, or subspecies, of modern man, or to a species different from *Homo sapiens*, or even to a hominid genus different from the genus *Homo*, of a kind perhaps already known as a fossil but not known to have survived to this day.

Of all these hypotheses, the first three turned out to be absolutely or practically impossible; the fourth can be eliminated pending further confirmation; and the fifth is clearly untenable. The only one remaining is that Hansen's specimen is a still unknown form of hominid. There is a good likelihood that it might already be familiar to prehistorians, in which case, it is an escapee of the past; in other words, a living human fossil.

Chapter 3

WHAT IT MUST BE.
What did prehistoric men look like?

"The historian does to the past what the crystal ball reader does for the future, but
the witch faces possible exposure, while the historian doesn't."
— Paul Valéry

Would an anthropologist recognize one of our own remote ancestors, if he were
by chance to meet one?

This is a thought that has long preoccupied me, since those days when I was a
university student, suckling the milk of official science. The pictures offered then of
the past and of the evolution of life were presented with such assurance that one might
have thought they were photographic documents. I could not help asking myself:
What if an eminent specialist of prehistoric hominids came up face to face with one
the beings which he claims, rightly or wrongly, as an ancestor… would he be able to
immediately identify it as such?

I am almost certain that he could not. The way that paleoanthropologists imagine
humans, pre-humans, and sub-humans of long ago is based on such a web of
preconceived ideas, such a rickety scaffolding of self-supporting hypotheses, that
there is a good chance they would be completely flabbergasted by the real appearance
of their ancestors.

In support of this view, just compare the various reconstructions proposed
over the years for the same fossil species. While *Australopithecus robustus*, for
example, is usually represented as a rough bipedal gorilla, its direct predecessor,
Zinjanthropus, is shown quite differently by its discoverer, Louis Leakey; it looks like
some old microcephalic Mongol philosopher. That may perhaps be an extreme case
of scientific dissonance. However there are many other species of fossil hominids
whose appearance is still a subject of debate. Take for example the Pithecanthropes;
some picture them as long-legged apes with a spark of intelligence in their eyes;
others as moronic looking Asiatic types with a low forehead and prognathous jaw.

Nationality even often enters in that kind of reconstruction. The fossil men
of Mikhail Guerasimov, nicknamed the "discoverer of faces," almost always look
like Soviet citizens; his Neanderthal from La Chapelle-aux-Saints is reminiscent of
Tolstoy; his more modern Cro-Magnon and Combe-Chapelle men look like kolkhoz
peasants straight from *The General Line,* a 1929 Eisenstein film. Similarly, a human
family tree by the Americans Douglas Gorsline and George V. Kelvin, published
in a *Time-Life* book, gives the most archaic types, from the lower Pleistocene,

mugs that make them look like convicts from Alcatraz, while those from the mid-Pleistocene would easily pass for triumphant football players at the Rose or Orange Bowl. And finally, upper-Pleistocene men look like escapees from a Western; one can even tell the good guys from the bad. Cro-Magnon man (our nearest ancestor) is the handsome cowboy, fearless and courteous, or perhaps the law-enforcing sheriff. There he is, escorted by his traditional acolytes: his right hand man is one of the Skhoul Neanderthals, a husky brawler, fond of bath-tub gin, but with a heart of gold, assisted by Rhodesia Man, faithful to the death; brave Uncle Tom. The half-breeds, Solo Man from Java and Chinese Peking Man, are characterized as horse thieves, scoundrels always ready to shoot their enemies in the back.

Political and social opinions do not seem foreign to the idea that people have of fossil men. The images of our past, seen through the lens of the western world, Anglo-Saxon, Germanic, or Nordic, nearly always represented *Homo sapiens* as blond, tall, and noble; while the Neanderthal, the cave-man, is shown as short, hairy, and shifty looking. In the days of colonial France, the old Neanderthal of La Chapelle-aux-Saints was depicted by professor Boule and sculptor Joanny-Durand as some kind of doltish Senegalese irregular; later, in a more democratic era, he is shown, particularly by the American Maurice P. Coon, with the looks of an average Frenchman. One finally wonders why the same Neanderthal looks like a Prussian squire in the eyes of German anthropologist Wandel, or a Brooklyn Jewish money lender in those of the American McGregor; or why at the Field Museum of Chicago, he looks like one of Al Capone's body guards.

It would be very interesting to submit all the authors of these reconstructions to a Szondi test. Their personal sympathies or antagonisms may well have played a major role in their anthropological ideas. In any case, the nationalistic aspects of all these reconstructions should inspire distrust and should convince us that even the greatest anthropologist would have a hard time determining the identity of any fossil man he would chance to meet. That would probably be the least of his worries, since he would likely be convinced that it couldn't ever happen. Nevertheless, that's the problem that Sanderson and I were presented with when we came face to face with Hansen's hairy man.

At the time of our first examination, it was too early to come to conclusions as to the zoological identity of the specimen. As fossil primates are known only by their bones, only an examination of the skeleton, with careful comparisons, could help determine if the creature could be identified with some supposedly extinct form, and if so, which one it could be. It was also not impossible that it could belong to some completely unknown species, even to paleontologists, although that was not likely.

In any case, based on what could be seen of the frozen man, it was already possible to hazard some guesses. What kind of fossil form could it be compared to? What forms could be surely be eliminated? What range of candidates could legitimately be considered?

While the specimen was a strange looking bird, he could obviously only be a mammal; its fur and breasts were clear evidence thereof. That within the class of Mammals he belonged to the order of Primates was equally manifest, although it was difficult to say exactly which criteria justified that choice. Truly, it was obvious at first sight because he clearly belonged to a mammalian sub-order, the Anthropoids, which includes apes and humans, as evidenced by its forward facing eye-sockets. Among Anthropoids, it clearly fit within those of the Old World, or Catarrhine monkeys, with nostrils opening forward and below (in contrast to New World monkeys, whose nostrils open to the side). Among Catarrhines, it belongs to the super-family Hominoidea, which includes both anthropoid apes and humans, although again it is difficult to specify exact criteria. Certainly the fact that it has human lower extremities, namely the feet of a plantigrade runner with a non-opposable big toe, is an argument in favor of that classification (that is not the case for anthropoid apes, which have feet with opposable big toes: feet that look like hands, especially adapted for living in trees).

One may point out that since the specimen in question has the feet of a human, there is no point in trying to identify it by gradually focusing on its zoological classification. Since it has the feet of a man, it must be a hominid.

But it's not so simple. There are completely different animals, bears for example, which also have plantigrade feet without opposable big toes. Besides if today anthropoid apes and humans clearly differ from each other through the opposability or non-opposability of their big toe, that might not always have been the case through geological history. Although that might be improbable, it is nevertheless possible that some day one might discover a terrestrial primate possessing most of the traits of anthropoid apes, but without opposable big toes.

That is why it would be most prudent to broaden the range of possible candidates to the entire super-family of Hominoidea.

So what Hominoids do we know today, either through observation or through the study of fossil remains? Furthermore, since we wish to discover which Hominoid most resembles Hansen's hairy man, we should also ask: What do the various Hominoids look like? Or looked like, if they have disappeared? Or might even look like if they have not yet been discovered?

To make things extremely simple, let's just say that we can distinguish among Hominoids six main types:[8] *Oreopithecus*, an ancient fossil with debatable connections;

---

8   Nowadays, the word "type" has acquired a bad reputation in anthropology, and rightly so; anthropology has become a "dynamic" science, in contrast to "typological." It studies fluctuating populations, changing nearly randomly under external as well as internal influences, rather than focusing around pre-conceived ideal "types." It is, however, on purpose that I use that term, in the sense of "a statistically representative mean of a population." This is no longer a type defined *a priori*, but *a posteriori*. In this way, I can gather under a single word a widely variable population, which I couldn't do if I used more precisely defined zoological terms like race, species, genus, or family.

the *anthropoid ape*, a true tailless ape; *Australopithecus*, a kind of bipedal anthropoid ape without fangs; *Pithecanthropus*, a kind of human with many simian features; Neanderthal, a human differing from us in a few simian features; and finally, *Modern Man*.

These six types have different levels of zoological importance. Within today's [1974] general classification adopted following the 1962 Burg Watenstein symposium organized by the Wenner-Gren Foundation for anthropological research, *Oreopithecus* by itself makes up a specific family; the anthropoid ape is represented by two families: the gibbons (or Hylogatids) and the great anthropoid apes (gorillas, chimpanzees, orangutans, i.e. the Pongids); Australopithecus is a family of its own, with only one species, *Australopithecus*; *Pithecanthropus* was formerly represented by many genera (*Pithecanthropus, Sinopithecus, Atlanthropus*...) but is now considered a mere subdivision of the genus *Homo*, the species *Homo erectus*; as to Neanderthal, it is now only a subspecies, *Homo sapiens neanderthalensis*; modern man, *Homo sapiens sapiens,* is another subspecies.

Let's forget about *Oreopithecus*, apparently neither anthropoid ape nor human, and probably the ancestor of neither. In any case, neither its anatomy nor its stature corresponds to that of our specimen.

Of the five remaining types, only two (at least in the eyes of official science) are still extant today: the anthropoid apes, of which we know a dozen species, and Modern Man, which belongs to a single, although highly diversified, species. The other types hark back to past periods: Neanderthal is rather recent; *Pithecanthropus* is older; and *Australopithecus* is even more ancient. An oversimplified view puts *Australopithecus* within the lower Pleistocene, *Pithecanthropus* in the middle Pleistocene, and Neanderthal in the late Pleistocene.

Since Darwin, scientific theory holds that man is the descendant of apes, and it was quite natural to link these five types to each other, each of the intermediate types representing a "missing link" eagerly sought at the dawn of the transformist revolution. The linkage from one type to another was first imagined in a rather simplistic way: each type morphed gradually over the millennia into the next one. The anthropoid ape gradually lost its fangs and adopted a bipedal stance to become *Australopithecus*; the latter stood up straighter, became more refined and cerebral, and became *Pithecanthropus*, which in turn gradually lost its simian features to transform into a coarse human: Neanderthal. And, finally, Neanderthal gradually lost its last brutish features to become the gentleman that we are.

All this was quite charming but stood in such contradiction with the chronological data and the tenets of comparative anatomy—the presumed descendants were often contemporaneous with or even older than their ancestors, who were generally more specialized in certain ways than their descendants—that it became necessary to bring serious changes to this rather simplistic and naïve version of human origins.

It was also realized that there was no direct transformation from one type to another but rather a differentiation among a population of creatures that led to a splitting into different daughter-populations that could replace each other and even coexist for long periods of time.

The original evolutionary chain was gradually augmented by lateral links to represent species, which after careful thought could not really be direct ancestors of Modern Man. So many such side chains had to be added, with secondary branches, that the original chain began to look like a tree. The idea of a chain of descendants was replaced by a family tree.

In the tree so constructed, the trunk represents the ancient ancestral stock of primitive forms, archaic and undifferentiated, while the upper branches stand for the more evolved and more specialized forms, those that have survived until today. Between those, within the foliage, some truncated branches represent extinct species that have not left any offspring. Overall, one could distinguish within that tree a number of different levels, and one naturally came to assimilate those levels to the types mentioned earlier, and to consider them as successive evolutionary phases that, through the ages, would have led from ape to man. The idea of a family tree morphed into a staircase.

This representation embodies real progress towards understanding the complex mechanisms of biological evolution, but basically, it all comes to the same thing. Whether one speaks of links or tree or staircase makes little difference.

Whatever evolutionary process is called forth, and whatever its symbolic representation might be, one is always harking back to Darwin's idea of a sequence of types leading from ape to man. And it is that classical sequence one must accept if one is to understand how to proceed to the reconstructions of the appearance of types that have (or are believed to have) disappeared.

Of course, of all Hominoids, we know only the appearance of *today's* anthropoid apes and of *today's* humans. We can't know very much about the appearance of humans, pre-humans, sub-humans, and anthropoid apes of yesteryears, of which we only possess rather incomplete skeletons, usually only skulls, sometimes part of a jaw, or the occipital bone, perhaps a few teeth... That has never kept artists and anthropologists from drawing numerous reconstructions of bygone forms, images that still haunt most zoology and physical anthropology textbooks.

After having been reproduced, copied, embellished, and improved along the same preconceived ideas, some of these images have taken on a traditional status that translates into solid belief. For example, today, Neanderthal Man is always represented with a white skin and only slightly hairy... actually less hairy than many normal people are today. He is usually shown with a full head of hair, bushy eyebrows, thick lips and hairy chin. At the other end of the scale, the Australopithecines are currently invariably depicted as chimpanzees or gorillas standing upright like people. It is only when we come to *Pithecanthropus*, whose nature has long been a subject of debate, that opinions diverge. However, as I pointed out at the beginning of this chapter, disagreement used to be generally much stronger. Why?

It is the nearly religious respect for the Darwinian sequence that has nearly always ruled the reconstruction of intermediate types, imagined by simple interpolation. In each representation, the relative contribution of simian and human traits has always depended on the personal opinion held by the author of the family tree of Hominoids and the degree of kinship between types. As this opinion was necessarily based on the knowledge of the time, reconstructions followed definite fashions. But, as Jean Cocteau put it, fashion is what becomes out of fashion.

Thus, when the only generally recognized "missing link" was Neanderthal man, the illustrator Kupka did not hesitate in picturing it as bipedal chimp with only the nose, lips, and feet of a human being. When the skullcap of the Trinil *Pithecanthropus* was recognized as that of a hominid, Neanderthal man began to look more human; it was the new type (Pithecanthrope) that had replaced it as "Ape-Man" (as its name indicates) halfway between the extremes. Later, after the discovery of the first *Australopithecus*, which took its place, the Javanese *Pithecanthropus* became more human-like, especially when its Chinese cousin *Sinanthropus* was found to be a toolmaker. A further reshuffling took place when it was discovered that there were two species of *Australopithecus*, the larger *Australopithecus robustus,* a vegetarian, and the gracile *Australopithecus,* a carnivore, like man. All the above now had to fit within the space between the ancestral anthropoid ape, *Dryopithecus*, and modern man. The increasing number of types saw them get more like each other.

Overall, within the currently known series of types, now including no less than fifteen distinct forms, one notices in the reconstructions that simian characteristics gradually trend to be replaced by human traits. One still gets the impression of actual continuity, but it remains an illusion and often contradicting anatomical facts. Nearly everyone agrees that these various forms are not derived directly from each other. Nevertheless, it is widely suggested that there appear to be successive waves of evolution. Few anthropologists can resist the temptation of showing a parade of types moving forward, rising from Ape to Man.

This kind of progression has the charming simplicity of cartoons where one can see, at the wave of the magic wand, Cinderella's pumpkin transform into a carriage and rats morph into liveried valets. However, they should not be taken any more seriously than those fairy tales.

One could even ask whether the famous Darwinian scheme according to which man descends from the ape might not also be such a fable, a fairy tale for scientists. After all, what is its factual basis?

I hasten to say that I am not in any way trying to minimize the impact of Darwin's extraordinary genius, nor debate the validity of his theory of evolution. The only issue is to assess the relevance of a detail: its application to the problem of our own origins. Of course, that minor detail is of major importance to us. But, since it stirs in us such deep feelings, it is not surprising that it would be handled with a certain degree of prejudice and with a level of passion foreign to the dispassionate and objective attitude of true science. Might it not be that applying the theory of evolution to our own origins could have been fundamentally flawed by incorrect ideas, a misinterpretation of the facts, or an outmoded view of the worlds?

The Darwinian conception of human origins is based on three ideas as old as History itself:

(1) Living beings are part of a natural gradation from the simplest to the more complex, in other words, a continuous chain of beings;

(2) Man occupies a privileged place in nature, at the highest level;

(3) Of all animals, apes are most like Man and thus occupy in the hierarchy of beings the level immediately below ours.

Combining these three ideas, apparently so obvious that they are treated as postulates, lead to unavoidable conclusions. Since Man, held to be the "King of Creation" or the "Apex of Evolution," stands in the first rank in the *Scala naturae* at the apex, it must be the final link in the chain of organisms. This is an idea that was already put forward by Persian Philosopher Nizami Aruzi Samarqandi in the 12th century. One also finds a similar interpretation in pre-Columbian America expressed in an Ecuadorian fresco.[9]

Incan fresco from Ecuador, representing the evolution of living beings.

9   On this particular chain of beings, namely a rope, life begins from a magma that organizes itself into spherical cells. These then stretch and become worm-like. The worm grows into a snake, acquires legs, and morphs into a crocodile, then into a turtle, which then loses its scales and begins to look like a mammalian quadruped. The latter then metamorphoses into a monkey, which soon loses its tail and takes a human appearance. The anthropoid then becomes a man, first naked and weaponless, but soon making bows and arrows. Then, all that's left for man to complete its evolution is to sprout wings and become some kind of angel. Teilhard de Chardin didn't invent anything!

When Science, starting in the 17th century, began to discover anthropoid apes and noticed that, of all monkeys, they were obviously the closest to man, they were naturally placed as the second-to-last link in the chain. The English physician and anatomist Edward Tyson was the first, in 1699, to speak of the chimpanzee, which he called "the Pygmy," as a link between monkeys and men.

The idea of a gradation of living beings was clearly widely accepted well before the theory of evolution. What evolution brought as a new concept was that these many links in the chain of beings derived from each other. The idea of transformism only provided an explanation for the similarities that linked all organisms.

The very idea of a transformation, of metamorphosis of one being into another, was not new in the days of Lamarck and Darwin; it was however only a popular fantasy. The belief that Man descends from Apes had been accepted since time immemorial by the Tibetans, as well as by the Bataks of Thailand, the natives of northeast Celebes, and the Yucarares of Bolivia. Darwin's original idea rested on his providing a rational explanation for such a transformation: natural selection.

In Darwin's days, the difference between anthropoidal apes and man seemed to be much too large to think of the former as the immediate ancestors of the latter. It was natural to suppose that the intermediate stage had disappeared and it was called "the missing link."

As we now know, it was not a single missing link but a whole series that were to be discovered. We are no longer missing any "missing links." One after the other, the fossilized remains of a plethora of creatures, which combined anthropoid ape and human traits, were unearthed. They were naturally fitted into a series of intermediate links that seemed to confirm the gradual transformation of an anthropoid ape into Man. If one blindly accepts the three master ideas listed above, the conclusion is irrefutable. But what are these ideas really worth?

The concept of a chain of living creatures is by now, as we know, an outmoded idea, long since abandoned. One knows perfectly well that the differentiation and speciation of animals has never been linear, like a chain, but has proceeded as in an extremely complex "bushy" family tree.

As to the idea that Man occupies the highest level in nature, it is an old western superstition as conceited as it is foolish. While the human species might be the first in terms of the development and complexity of its brain, that is certainly not the case for a multitude of other characters. Unaided, Man cannot flit in the air like a bird, nor navigate in the dark like a bat; he cannot swim as fast as a tuna or a dolphin, nor dive into the abyss like the sperm whale; he cannot kill with a single bite as some snakes can; he cannot regenerate its limbs, like an amphibian; he cannot make itself nearly invisible like chameleons; he is incapable of sinking into apparent death as can many invertebrates. No animal is absolutely superior to all the others; each one is superbly adapted to its environment and is the king of its own ecological niche. Of course, it is because of his intelligence that Man has become master of the world; it is also because of the way he has used this great intellect that he has gradually destroyed or poisoned his environment, gradually leading to his own demise. Man is not even capable of using wisely what constitutes his main claim to glory.

The only one of the above trio of ideas that remains defensible is the close resemblance between Man and Apes. Not only has it survived the tides of scientific enquiry, but the most advanced research in comparative anatomy, physiology, genetics, and biochemistry have confirmed that the similarity reflects true kinship. Research has determined without doubt through the structure of their hemoglobin, the nature of blood proteins, and the shape and number of chromosomes, that humans and the great African anthropoid apes (gorillas and chimpanzees) are closely related.

That two beings are related does not necessarily mean that they are father and son, or grandfather and grandson. They might be uncle and nephew, or great-uncle and great-nephew, or cousins. Furthermore, one has to find out which is the father or grandfather, or the uncle, or great-uncle of the other. It is obviously the older of the two, one might say. That is undoubtedly so in a line of direct descent. But it is quite possible to be older than one's uncle. In matters of paleontology, one rarely deals with direct lineages, and it is often difficult to be sure as to which of two species is most ancient.

The falsity of two of the three basic ideas upon which originally rested the traditional conception of human origins changes everything in the appearance of human family tree. Anthropologists and primatologists seem to have forgotten, or deliberately ignored this fact.

Since there is no simple linear link of animal species, but rather a bushy family tree, and since Man is neither an end point nor an apex, as its pride would suggest, and that he does not necessarily stand *beyond* other Hominoid species, why should one faithfully stick to the old Darwinian sequence? It is perfectly legitimate to wonder whether anthropoid apes might descend from man rather than the opposite.

In order to discover the direction in which evolution proceeds, there exist two series of reference points: first the sequence of fossil forms through geological ages; second, the development of individuals as they grow towards adulthood.

Although, at first, paleontology gave the illusion of providing an answer in conformity with the Darwinian scheme, a plethora of further discoveries have demonstrated that the various fossils which were thought to be linked to each other had coexisted for long periods of time, or had even followed each other in the opposite order, so that the various phases they were supposed to represent quite embarrassingly overlapped.[10]

Embryology provides more reliable evidence. The great German biologist Karl

---

10   The more conscientious and lucid paleontologists are finally yielding to the evidence; the old Darwinian scheme is less and less compatible with scientific data. One should consult a recent work: *Inte från aporna* (*The Monkey's Step*) by a prominent Finnish paleontologist Björn Kurtén of Helsinki University. An English translation, *Not from the Apes,* was published in 1972.

von Baer noted in the middle on the 19th century that "the younger the embryos of various animals, the more they resemble each other, and the older they are, the more they differ." It is on that basis that Ernst Haeckel stated his famous biogenetic law according to which ontogenesis recapitulates phylogenesis. In other words, the embryonic development of each individual reproduces over a shorter time the evolution of its species.

Since it was first stated in 1860, the law of recapitulation of ancestral stages has been the subject of extensive criticism and has been somewhat modified. Whether one accepts it verbatim as expressed by Haeckel, or whether one prefers, more wisely, to follow the views of the Russian Severtsov or the English Beer that ontogenesis recapitulates the ancestral *embryonic* stages, one must accept that embryonic development shows unambiguously *the direction in which evolution has taken place*. So, what do we find?

Comparing embryonic and even post-embryonic development in Man with that of anthropoid apes, one notices that while apes go through a stage that is strongly reminiscent of Man (particularly in the globular shape of the head, a high forehead, the absence of prognathism), Man never goes through a simian stage. That is, of course, extremely embarrassing for the devotees of the Darwinian scheme, and it is not surprising that the law of recapitulation has been vigorously contested.

De-hominization of the chimpanzee, from infancy to adulthood.

In spite of that, one doesn't have to turn Darwin's scheme upside down and declare that Man descends from Ape. That would simply mean going from one excess to another. We already know anyway that the known hominoids are not part of a single chain of descent and could belong to different branches or twigs of the super-family's ancestral tree.

Today, everyone agrees that Man and the existing anthropoid apes are more or less remote cousins and that their remote common ancestor in the Tertiary was certainly not some kind of archaic pongid ape like *Dryopithecus*. It had to be more like Man than like a brachiating ape. It was probably some kind of infra-pygmy, a round-headed gnome, walking upright, in other words, the *Eoanthropus* imagined by leading anthropologists such as Marcellin Boule in France and Henry F. Osborn in the USA.[11] Although their opponents, led by William King Gregory, the pope of American paleontology, have dubbed them Homonculus-dreamers, each new paleontological discovery forces us to pay more respect and attention to their speculation.

The remains of hominids are continuously discovered from ever-older geological strata. In the Omo valley of Ethiopia, teeth with human characteristics have been found and dated at 3,750,000 years, in the Pliocene era; also, minuscule quartz tools at least two million years old, dating from the upper Pleistocene. In 1967, when the great Louis Leakey allowed me to handle the freshly reconstructed skull of *Homo abilis,* a small bipedal hominid with a rounded skull, I could not avoid thinking of the *Eoanthropus* imagined by the famous French anthropologist and the no less reputed American zoologist. The small *Ramapithecus* from India and Kenya, which dates back twelve to fourteen million years, at the end of the Miocene, is currently known only through a couple of upper jawbone bearing teeth arranged in a semi-circle. If some day, thanks to the discovery of a more complete skeleton, it were to turn out to be what it is suspected to be, namely a hominid, and thus a miniature man, the hypothesis of the Homonculists would triumph.

In spite of the rather long digression required, this crucial point had to be made in order provide a better perspective of what past and present hominids could have looked like, to help us better understand the characteristics of our specimen, and to situate it more precisely with respect to modern man and to the various fossil hominids and anthropoid apes.

Now that we know that reconstructions, by mere interpolation, of forms of controversial ancestry cannot be of any value, we can quietly ask the question: What can we really know of the appearance of a creature of whom we have only bones?

Starting with the skeleton, which bears traces of the placement of the tendons and of the size of some muscles, it is possible to reconstruct with some precision the musculature. However, according to current scientific opinions, that is as far as rigorous deductions can lead us.[12]

Mikhail Gerasimov does not agree. Specifically, he claims that the face of a

---

11   "What we need as an ideal common ancestor," according to professor William Howells, "is a generalized non-leaping tarsier, a yet undiscovered creature, the dream of optimistic paleontologists."

12   On this problem, see McGregor (1926), Loth (1936), Schultz (1955) and Kurth (1958).

hominid, in particular the shape of the nose, mouth, eyes, and ears can be reconstructed from a study of the details of the skull. He devoted his whole life to prove it, with great success, at least on modern men. It is astounding how well he managed to reproduce the appearance of some people based only on their skull, with a degree of precision that could be verified from portraits or photographs. The resemblance is so striking that in some cases his work assisted some forensic enquiries.

Gerasimov's extraordinary skill, which led him to be called "the discoverer of faces," was based on a careful and exhaustive study of a multitude of specimens. His studies allowed him to establish some links between various skeletal forms and structures and the shape of the soft tissues covering them, taking into account age, gender, and racial factors. So, even though he undertook the reconstruction of some 200 prehistoric hominids, the great Soviet researcher readily admitted that for those "the norms established for modern men no longer sufficed." Furthermore, although he recognized that the classical Darwinian sequence leading from ape to man was unacceptable, he nevertheless fell under its spell. When reconstructing an Australopithecine, he naturally gave it the appearance of a chimpanzee.

It is obvious that Gerasimov's techniques are not fully transferable from Modern Man to beings whose anatomy is significantly different. One might even wonder if his techniques would have as much success outside the Soviet Union. Most of the corpses on which "the discoverer of faces" worked to establish his norms were his own compatriots, and it's not surprising that his Cro-Magnon man should have a Slavic look. And that precisely because of, rather than in spite of, his rigorous methodology.

As soon as the architecture of the skull differs, it's clear that some details of the nose, lips, eyes, and ears can only be invented. Furthermore, what could one possibly know of the flesh, the skin color, and the arrangement, length, color, and texture of the hair? If you think about it, all the different great cats—lion, tiger, leopard, jaguar, snow leopard—have just about the same skeleton, which is why they are classified within the same genus. If we knew them only through their bones, we could not tell the color of their skin, or whether it was of a uniform hue or striped or mottled, and whether they had a mane or not. For example, we don't know whether what we call the cave lion was not perhaps a giant tiger or leopard.

Bones alone would not tell us that the pangolin has scales, or that the hippo is hairless, but the mammoth is wooly. And we could only speculate as to the shape of their eyes, ears, or nostrils.

Our degree of ignorance of fossil humans can't be as great, but might approach it.

Nothing proves, for example, that Neanderthals, our nearest relatives, were not as hairy as orangutans. Nothing suggests that they might have long hair, rather than short hair on their head, or that they might have had the beard of a prophet or a pop singer, rather than the naked face typical of most apes. Nothing allows us to decide that they had the thick lips of Negroes rather than the absence of lips as in apes. It is not certain at all that they had white skin and black hair. And we are certainly not sure that one could see the white of their eyes, as with modern man, rather than round eyes showing only the iris, as anthropoid apes have.

As we now possess more than a hundred skeletons of Neanderthals, some nearly complete, we know their morphology in great detail. It is believed that the most characteristic among them, the most "classical" as they are called, had a disproportionately enormous face, flat but projecting forward like a snout.[13] Their forehead sloped back behind heavy eyebrows reaching completely across, somewhat like a visor. They had a heavy jaw but no chin, and their rather elongated head ended at the back with a kind of a bun. They were barrel-chested, had short and thick forearms, large hands, bowed femurs, relatively short lower legs, and very large feet with splayed toes curving downwards. It is thought that they carried their head bent forwards, deep between their shoulders and held by a bull-like neck, and that they walked with bent knees. That is how much one can say about their general appearance, although that is still a matter of controversy.

Over the past few decades, the general tendency among anthropologists, especially North Americans, has been to de-emphasize as much as possible the simian appearance traditionally attributed to Neanderthals and by now solidly anchored in popular imagination. This change of attitude is due to the fact that the description of these creatures was mainly based at the beginning on the specimen from La Chapelle-aux-Saints, the best preserved of all, but thought today as belonging to a rheumatic oldster heavy with age and of a somewhat extreme, even aberrant, type. As summarized by C. Loring Brace, "From what we now know, it's likely that if a Neanderthal was shaved and properly dressed and mixed with a crowd of people shopping or catching the subway, he might look perhaps slightly unusual—short and stocky with a prominent jaw—but nothing else." Clark Howell went a step further: "Dress him in a Brooks Brothers suit and send him out for groceries at the supermarket and he would probably go unnoticed."

These views from two prominent American anthropologists are likely exaggerated but a good reminder of the lack of unanimity of our ideas about the appearance of our Neanderthal uncles and of the depth of our ignorance.

And what about the even lesser known Pithecanthropes? Besides about twenty more or less complete skulls, but differing from each other according to their origin, only bone fragments, mostly femurs, have been discovered. In view of their aggressive prognathism and their sloping forehead, these Archanthropians, as they

---

13   This facial structure, which Prof. Sergio Sergi has dubbed oncognathism, is not to be confused with prognathism, in which only the jaws protrude forward. Apes are prognathous; Neanderthals are oncognathous.

are sometimes called, must have had much more simian faces than Neanderthals. However their general appearance might not be quite as unfamiliar: their legs were as long and straight as those of modern man, and so presumably was their gait.

As to the Australopithecines, nothing can be said about their appearance. There are only a few of them, and other than a few skulls belonging, as we know, to two distinct forms, they are known only from three or four fragments of pelvis, a femur and a tibia, and a number of small hand and foot bones. That hasn't kept everybody from representing them as bipedal anthropoid apes, well on their way to human-hood. However, in November 1971, Richard Leakey, the very son of Louis and Mary Leakey, announced that on the east side of Lake Rudolf, in Kenya, he had dug up numerous remains of Australopithecines, including a whole assortment of limbs. An initial examination of this great discovery was to show that, in fact, the Australopithecines had long arms and short legs and probably walked by leaning on their knuckles, just like gorillas and chimpanzees. This news dismayed traditional anthropologists, many of whom blamed their younger colleague for such a shocking announcement.

In the hairy man preserved by Hansen, we finally find ourselves faced with a complete specimen of a hominid presenting both human and simian characters. It was certainly neither an anthropoid ape nor a modern human. For lack of other options, it was a creature to be classified among the three so-called intermediate categories: *Australopithecus*, *Pithecanthropus*, or Neanderthal. For which one of these would we now be able to fill the gaps in our anatomical knowledge? It did not seem possible at first sight to be sure. Whatever it was to be, its study would certainly throw light on the evolution and the differentiation of hominoid primates and, most important, on our own origins.

In order to venture an opinion on the identity of this unique specimen, one should be influenced neither by its superficial simian characteristics nor by even the best reconstructions proposed for the various forms of fossil hominoids; nothing but its own morphology should be taken into account.

It was nevertheless reasonable to first ask oneself whether this contemporary creature might not be identified with the *most recent* fossil form. After all, while the Australopithecines are presumed extinct for 400,000 years and, but for a few exceptions, the Pithecanthropes have been gone for some 150,000 years, Neanderthals have been missing only for 50,000 years. Wouldn't that make them the most likely candidates to the title of "living fossils" of the human genus?

Most of the anatomical characteristics of Hansen's hairy man fit with what we know of *classical* Neanderthals. Namely, the width of the face, very flat; a sloping forehead behind thick eyebrows forming a ridge; the heavy lower jaw and absence of chin; the head sunken into the shoulders; arched collar bones; a barrel chest; thick and short forearms and lower legs; enormous hands; surprising wide feet and widely spread hook-like toes.[14]

The only feature of the specimen that did not seem to fit with the preliminary identification was its stature. Classical western Neanderthals were on the whole massive and stocky, generally no more than 1.65 m [5 ft 5 in] tall. But that was not so for all Neanderthals: specimen I of Chanidar, in Iraq, measured between 1.70 and 1.73 m [5 ft 7 in and 5 ft 8 in]; specimens IV and V from Skhoul in Palestine reached 1.70 and 1.80 m [5 ft 11 in] respectively. Anyway, if Hansen's specimen was a Neanderthal, it was a later one, a member of a lineage that had continued to evolve over 50,000 years. An increase in size is one of the most common trends in evolution. Consanguinity within small isolated populations would have accelerated such a trend.

But might the observed features of the specimen agree just as well with those of *Pithecanthropus*, or even of the Australopithecines? Apparently not. In both of these forms, and especially the second, being known today mainly from skulls, it is basically from cranial characteristics that one should judge whether the specimen belonged to one of those forms. But in fact, the shape of the Pithecanthrope's face, the width of the jaws and its pronounced prognathism, a feature even more prominent among Australopithecines, especially among the carnivorous types, are incompatible with the appearance of the hairy frozen man.

So, if Hansen's specimen did not belong to a species of hominids still completely unknown to anthropologists and paleontologists, *it was most likely to be a belated Neanderthal*. If that was indeed the case, one had to admit that it did not at all resemble, at least superficially, most of the reconstructions of early man, especially the most recent. Clad in an embroidered jellaba, he might easily go unnoticed as a musician in a rock band, but even wearing a Brooks Brothers suit he might not gain entry into Polite Society.

This terrifying hairy monster was obviously not the only one of his species and its congeners must have spread around the Earth since prehistoric times. One might well ask: "How come we have never seen any?" or alternately: "We would know about it!"

Such an attitude demonstrates a deep ignorance. The fact is that hairy wild men have been seen and described since remote times and continue to be seen today. So much was well known. I had catalogued them in one of my own books. Sanderson had published a synthesis of what is known about them. Porshnev, after a thorough investigation, had identified their most common appearance as a form of relic Neanderthal. And now it was clearly a representative of this type that we were looking at.

---

14   After devoting two years to an in-depth study of the external anatomy of the specimen from photos, my views were confirmed and I also discovered more subtle as well as specific traits linked to the upturned form of the nose, the length of the thumb, as well as its weak opposability.

Over the first few hours of my examination of the specimen, I had caught myself thinking: "Porshnev was right! Porshnev was right!" At first, it was like a flash of inspiration, a blinding truth from within that one's analytical mind couldn't immediately accept. That intuition was no doubt but the echo of the argument elaborated in this chapter that had flashed like a spark in my mind. However, as I had in hand more and more relevant traits, and in spite of a few surprising features, I became increasingly certain and eventually absolutely sure: "Porshnev *was* right. Neanderthals have survived to this day. This is one of them."

So, for a year and a half, crowds had marched past the frozen coffin where rested the remains of a Neanderthal. There must have been among them some physicians, or biologists. No one, apparently, had realized the incredible scientific importance of this specimen. At least no one had dared mention it in a scientific publication.

True, in order to "discover" this creature and appreciate its significance, it was necessary to be a researcher expert in physical anthropology and comparative anatomy and keenly aware of the state of knowledge on hairy bipeds still unknown to science. Nobody in the world was in a better position in that respect than Sanderson and myself. And luckily, there we were, both of us!

Ivan had been mocked by many establishment scientists because of his heretical views on zoological mysteries and in particular on the Abominable Snowman; his reputation as an excellent naturalist had gradually been eroded over the past few years to the point that, in the eyes of some, he was a mere scribbler for cheap magazines; Ivan, the pariah of science, was finally going to take his revenge on his critics. As for myself, my knowledge of one of my "unknown beasts"—an "unknown man" in this case—was about to enter a new phase: cryptozoology was bearing its fruits, and what fruits! The specimen that everyone ardently wished for was finally at hand, and its peculiar anatomy was undisputed evidence of its authenticity. Nobody could, or at least should, be able to deny its existence. I would be able to write a detailed description to be published, with photos and illustrations, in a reputable scientific publication, the *Bulletin of the Institut Royal des Sciences Naturelles de Belgique*, with whom I had close ties. The whole world would speak of our fantastic discovery! Many colleagues, zoologists and anthropologists, would express their enthusiasm.

Well, all this was indeed to happen, but one should not imagine that such a discovery, in spite of its strong scientific underpinnings, would be recognized and unanimously accepted by Science… That would be to misunderstand that irrational and apparently incurable suspicion based on an instinctive fear of the unknown as well as the reluctance felt by the human mind to perturb its preconceived ideas and its intellectual comfort.

A fossil man is by definition a being that no longer inhabits the world. Anyone claiming to have seen one alive is either a liar or a hoaxer, drunk, or myopic. When anyone discovers bones in recent strata, which has often happened with ancient humans, it is immediately said that the ground is older than was thought, or that the human found was a modern man with archaic traits. If anyone is bold enough to come up with a corpse, it can only have been preserved since prehistory. Or it must be something else, perhaps a monster that only looks like a fossil man, or some clever artificial construction.

With a few rare exceptions, mostly open-minded colleagues or old friends, the reaction of people to whom I was to speak of our discovery, supported by photographs and my scientific report, went from laughter to cautious expectancy with, in–between, a shrug or a half-smile from those who wonder how much money this is worth to you or the instinctive withdrawal of someone who suddenly begins to doubt your sanity.

I often thought of my late friend Professor J.L.B. Smith, who lived through a similar experience. When I visited him in Grahamstown, South Africa, he told me how after his publication of a description of the first coelacanth in 1939, some of his colleagues thought he was crazy, and a number of friends and acquaintances had started avoiding him and stopped greeting him in the street. And that, with a specimen in his laboratory that everyone could come and touch.

I also thought of Miss Courtney-Latimer, who made the discovery, and whose name is immortalized in that of the famous four-legged fish *Latimeria chalumnae*. When I saw her at the East London Museum of Natural History, which she now heads, she told me that at the time, an eminent British paleontologist, W.E. Swinton, had called her to confirm that the fish had indeed been fished out of the mud. When she replied that this was certainly not the case, he had pressed her to say so to the press. Incapable of admitting the possible survival of the coelacanth, he wanted to convince the public that the fish had been preserved in mud for over 60 million years…

A fossil fish should remain fossilized!

The discovery, in our time, of a human thought to be prehistoric is, of course, even more shocking than that of a live coelacanth. One might reasonably think that it should appear more likely, since the famous four-legged fish was presumed to have been extinct for tens of millions of years, while Neanderthal man was only 50,000 years old. But scientists stop being reasonable when dealing with their close relatives. This business was much more relevant to us humans than the origin of land quadrupeds; it also played havoc with deep-rooted popular beliefs as well as with the basic dogmas of anthropology.

A fossil man will please remain a fossil!

So our "ape-man" faced a double threat of being returned to the fossil state that he should never have left. Most people simply thought of burying, at least figuratively, this flesh and bone anachronism. But worse, one could fear that those who possessed the specimen would actually bury it, for still unclear personal reasons.

For, as we shall see, the seamy side of this affair was even more mysterious.

Chapter 4

WHY SO MANY MYSTERIES?
The eloquent silences of a carny hawker.

"These are not natural events; they strengthen
From strange to stranger."
— William Shakespeare, *The Tempest*, Act 5, sc.1

When Ivan T. Sanderson and I had the opportunity to examine the hairy man whose corpse had been exhibited for eighteen months in American country fairs, we had been immediately struck by the ambiguous behavior of its exhibitor, Frank D. Hansen. He constantly contradicted himself and disputed what we could see with our own eyes. He eluded the simplest questions about the history of the hairy creature and offered different variants about its origin. He jealously guarded his specimen but was also trying to convince us that it was a clever fake or a worthless item without scientific value. In the final instance, he referred all decisions to a powerful Californian mogul, the "real owner," whose name he would not reveal.

The dramatic incidents that followed our visit only confirmed that there was in this affair "much more than meets the eye," as one rightly says in the US.

After spending three days studying, measuring, drawing, and photographing the specimen, Sanderson and I were absolutely convinced of its extraordinary value and had only one thought: how to submit it as soon as possible to a rigorous scientific enquiry. We had to either obtain permission to take an x-ray scan that would convince the most skeptical, after which scientific institutions could bid for the specimen, or find a wealthy person who would be willing to buy it sight unseen, on the faith of our conclusions.

Just in case, Sanderson, who was confident that he could raise the money among its well-heeled friends, asked Hansen for an option on the possible sale of the specimen. A substantial sum of money would be deposited in the bank as a guarantee in case a transaction became possible.

Charging 35 cents a head, Hansen claimed to have made $50,000 a year[15] and hoped to continue showing the specimen for another year to cover his costs and realize some profit. Sanderson pointed out that selling the specimen immediately would bring in much more than the $50,000 he was hoping for: at least twice if

---

15   As I shall show later by some calculations, that amount represents a theoretical maximum, nearly impossible to reach in practice.

not four times as much and perhaps more. He reminded him that in 1961, in New York, in front of many lawyers, he had heard agents for *Life* magazine offer the late Tom Slick the sum of half-a-million dollars for the first photo of an Abominable Snowman, dead or alive, provided scientific experts would give written confirmation of its authenticity. That much just for a photo! One could dream of how much the real flesh and bone specimen of an unknown hominid would be worth. One could expect at least twice as much…

As for myself, I thought it was a little clumsy to raise the ante on an item that we wanted to acquire, which could only raise its selling price. However, as time would tell, even a million dollars would not have carried the day.

Hansen merely said that he would forward our offer to the owner; he was later to tell us that the owner did not want to sell for *any amount.* The mysterious Californian nabob would invariably, and always through the intermediary of Hansen, present the same absolute refusal to all further offers, *even if, it should be noted, the identity of the specimen was not guaranteed.*

As Hansen was to state on the phone to Miss Marlene Simons of the *London Times*, in New York, on March 25, 1969, the mysterious owner was a very wealthy individual whose pleasure was "to own something very rare, something that nobody else had."

Apparently, that person was even more defensive of his possessions than we could imagine, for during our visit to his farm Hansen told us that he had spoken to him on the phone and had been reprimanded for having allowed us to examine his specimen. He added that he would henceforth let no one examine it. He was even thinking of cancelling all contracts already signed for further showings in Canada and the US. He was considering, he said ironically, "putting the whole thing on ice."

We were soon to see that this was no idle threat.

Once back in New Jersey, Ivan and I set on, independently so as not to influence each other, to write a preliminary report containing initial conclusions based on our own personal examination of the specimen. These two reports were the first draft of notes aimed at publication in scientific journals. I was planning to send mine to the Institut Royal des Sciences Naturelles de Belgique, of which I was and continue to be a member, and Ivan planned to send his to *Genus*, a sociology journal published in Rome by Professor Corrado Gini, wherein were usually published the works of the International Committee for the Study of the Humanlike Hairy Bipeds.

The time had come to give a name to the creature that we had examined. For simplicity, and because this is an American habit, Ivan had given him a code name: Bozo, which was also unfortunately that of the most famous clown in the US. That name seemed to me particularly unfortunate. I already knew that no occasion would be missed to throw ridicule upon the whole affair, as had already been done on the Snowman file. We should be careful not to provide ammunition to our detractors. This was a serious business, which had to be handled with all the gravity it deserved.

It was imperative, I thought, to give to this creature the scientific name it deserved. We had all the elements needed for a detailed and unambiguous description—a diagnosis, in zoological jargon—of the external anatomy of the hairy man. The specimen that we had studied could legitimately be considered as the holotype of a form of an as yet unknown hominoid, a form of which no complete and valid scientific description had been published.[16] This gap had to be filled.

What name could we legitimately give to the zoological form of which the frozen man was a representative? That he was a hominid there was no doubt, and as hominids are currently gathered in a single genus, he was a *Homo*. But of what species? For me, as I explained, it seemed to belong, through most of its characteristics, to the Neanderthal. At that time however, Ivan, for reasons he never explained, was more inclined towards the Pithecanthropes (*Homo erectus*), or even more archaic hominids.

If he was a Neanderthal, what should he be called? There exists among anthropologists a great degree of dissent as to the taxonomic status of that form. For some, it is actually a completely different species from *Homo sapiens*, *Homo neanderthalensis*; for others, it is only a subspecies: *Homo sapiens neanderthalensis*.

The first unambiguous description of a relic Neanderthal came from Professor Vitali A. Khakhlov. Although it was based on observations made between 1907 and 1915, it was not published until 1959 by Porshnev and Chmakov. Unfortunately for the prominent Russian zoologist, the name he chose, *Primihomo asiaticus*, cannot be valid; creatures to which it is applied are indubitably of the genus *Homo*, and the name *Homo asiaticus,* which would be more appropriate, had already been used by Linné to describe a subspecies of *Homo sapiens,* namely Asiatics.

However it is customary in zoological nomenclature to attribute to a relic form of a species thought extinct for millennia a status equivalent *at least* to that of a geographical race, namely that of a subspecies. A new name had to be found for the form represented by the frozen man.

"What if we called him *Homo pongo*?" proposed Sanderson. "Hairy as he is, he really looks like an anthropoid ape."

Pongo is the generic name of the orangutan, and it is the name of the family containing all great anthropoidal apes, the Pongidae.

"That's a great idea," I replied, "but it seems to me that we should not stick together two names in this manner. This would seem to suggest that we think of him

---

16   Although it is true, as Porshnev has pointed out, that Linné had already in 1758 given the name *Homo troglodytes* to the wild and hairy men he had often heard of, the great Swedish naturalist had not chosen his references well. The cases he cited were of people suffering from albinism and in one case most likely of hypertrichosis. The diagnosis of the species certainly applied to albinos: it spoke of small men with white skin, white curly hair, and pink eyes.

as an ape-man. But he is really a man; he only superficially looks like an ape. Why don't we call him *Homo pongoides*, i.e. an 'anthropoid-ape-like man?' Exactly as we call anthropoid an ape that looks like a man."

"Agreed." said Ivan. "It has a good ring to it. *Homo pongoides*, the pongoid man."

So a new name was born, one that we hoped would have great success.

However this baptism did not settle the matter of finding a scientific name for our specimen once and for all. If it was, as I believed, a Neanderthal, it deserved as a relic form the name *Homo neanderthalensis pongoides* in the eyes of those anthropologists concerned with Neanderthal specificity. But how would others call it? *Homo sapiens pongoides*? It wouldn't make sense to place the pongoid man in a category parallel to today's major human races. It was much too different.

In any case, it would be strictly legitimate to place our hairy man within the species or subspecies of known Neanderthals only when his bones could be compared with those of fossil specimens. Who knows? Perhaps it would turn out that pongoid men had evolved so far in the direction first taken by generalized Neanderthal types (the pre-Neanderthals) and followed by the more extreme types (the classical Neanderthals) that they would have reached the threshold of specificity. In that case, they would deserve being placed in a completely different species.

These issues were merely of academic interest. Having described a new form of hominoids unknown to science, and still living with us on Earth, as *Homo pongoides*, and prudently following it with the label *sp. seu subsp. nov* (meaning new species or subspecies), I was strictly following the rules of zoological nomenclature and felt beyond reasonable criticism.

Once that small anatomical detail was taken care of, I strove in my note to point out how the discovery that Sanderson and I had had the incredible luck to make was the tangible result of a long and dogged effort by a whole team of collaborators all over the world. After this acknowledgment of the work of all the pioneers, I felt justified to state, rather emphatically that:

> Those whom luck has favored wish to associate their success with their long-time friend and colleague Boris Porshnev, without whose significant contributions they could never had fully grasped the problem; in our discovery, he also sees the triumph of his own thesis.
>
> May this success, the result of a close international collaboration by three researchers, respectively from the United States of America, Western Europe, and the Soviet Union, throw a new light on the problem of the origin of the human species, and also contribute to a better understanding between the various people it comprises.

I wrote those concluding lines of my preliminary note on December 29, 1968, at

Ivan T. Sanderson's farm, near Blairstown in New Jersey.

Before claiming final victory, we had to set up a detailed plan to try and obtain from Hansen permission to perform a meticulous examination of his frozen specimen. Ivan and I spent the Christmas and New Year holiday planning our next move. It quickly turned out that we had very different views on how to gain access to the specimen.

Ivan, having lived in the United States for about twenty years, was in a much better position than I to take care of many practical issues. Authoritarian by character, nearly dictatorial one might say, he wanted to take sole and personal charge of "rescuing" the priceless specimen, either by crude commercial means or, if these were to fail, by an appeal to the coercive powers of the law.

As for me, I was convinced of the fundamental necessity of alerting the international scientific community to urge American public opinion *to take our discovery seriously*. Well aware of the public's skepticism when faced with anything new, unusual, or revolutionary, I did not delude myself of the difficulty of this task. But as long as that obstacle was not overcome, *nothing* was possible. How could one expect a scientific institution or some generous philanthropist to consider spending a fortune on a specimen of dubious nature?

And just how were the authorities eventually to be convinced to force Hansen or the anonymous owner to hand over the specimen?

As I had little flair for business and was quite allergic to legal and administrative questions, I would have been delighted as well as relieved to let Ivan take care of the whole affair. But I had not failed to notice that because of the rather undiscriminating interest he had shown over the past few years to everything that was fantastic and unexplained, the naturalist and author of many well-received works had gradually acquired a wretched scientific reputation. While acknowledging his originality, I had many times chided him in a friendly manner for the excessive enthusiasm with which he sometimes blindly welcomed marvelous events without verifying their authenticity. That had started to play against him. Nowadays, it was enough to mention the name Ivan T. Sanderson in some scientific or even publishing circles to see people smile and shrug. I found that very painful, for I appreciated the exceptional intelligence and skill of my old friend, but that was a fact, a fact which was to make our work in this delicate matter much more difficult.

Ivan was in favor of strong measures: he believed that the issue would be settled by the power of money or by that of the police. Perhaps he was right as to the effectiveness of those methods, but as a matter of principle, I had more confidence in wile and diplomacy. We ended up with a gentlemen's agreement: Ivan's iron fist would hide in my silk glove. I would be in charge of convincing scientific circles, or at least some influential scientists, to gain their moral support. Based on this support, Ivan would then take action, moving as many pawns as possible in this game of chess:

financiers and journalists, policemen and lawyers, bureaucrats and politicians.

All of which, of course, would remain a complete secret. We even agreed to adopt in our relations with Hansen and his patron a technique well known in police interrogations: the good cop, bad cop roles. I was to be the "bad-cop"—since I was leaving the US for an expedition to Central America, after which I would be returning to Europe, I could be blamed for all disagreeable moves. Ivan, on the other hand, would strive to stay on the best of terms with Hansen and even give him the impression of siding with him against me.

After completing our reports and having had the photographs of pongoid man developed—they turned out very well—Ivan and I decided to go and visit together one of the most eminent anthropologists in the United States, Dr. Carlton S. Coon. We absolutely needed his opinion on the best way to bring the specimen in for scientific analysis and hoped also to gain his support.

We were back on the road on January 2, 1969. After a night in New York, and picking up in Connecticut our friend the geologist Jack A. Ullrich, we continued towards Massachusetts where Dr. Coon welcomed us on the 4th at his home in Gloucester. He had spent part of the night reading our reports, sent to him the previous evening, and after scrutinizing enlargements of my photographs and projecting my colored slides, he finally declared: "I agree with you on three points: it is an authentic specimen, there is no doubt about it; it is something unknown to us; and it is to be classified among hominids." But he went on with a tired smile: "I say this between us, but I will not compromise myself by making a public statement. I already have enough trouble with my own work!"[17]

Stimulated by the complementary and concurring diagnostic of such an expert, while also disappointed by his refusal to publicly support us, we returned to Connecticut, where I was to prepare, as a guest of Jack Ullrich, our joint expedition to Central America, and where I took leave of Ivan Sanderson. Henceforth, we were to work and act completely independently of each other.

In Westport, at Jack's house, I mainly kept myself busy with finalizing my preliminary note and its illustrations. By putting together four enlargements of the color photos of the frozen creature taken from the same angle and at the same distance, I managed to obtain an overview of its general anatomy. From that composite photo and with the help of close-ups and sketches, I put together a rather thorough drawing

---

17   Dr. Coon was alluding to the passionate reactions to his book *The Origin of Races* and to the abusive political interpretations that followed. I would not have repeated here the confidential words of the great anthropologist if he had not himself changed his mind and authorized Ivan to publish his views in *Argosy* in May 1969. "The photos and description of this specimen demonstrate that it is a complete corpse, and neither a composite creation or a fabricated model. In addition, it is not just a hominid, but a kind of man, although it exhibits a number of unexpected anatomical traits which will be of the utmost interest to physical anthropologists."

of what the pongoid man would look like once thawed.

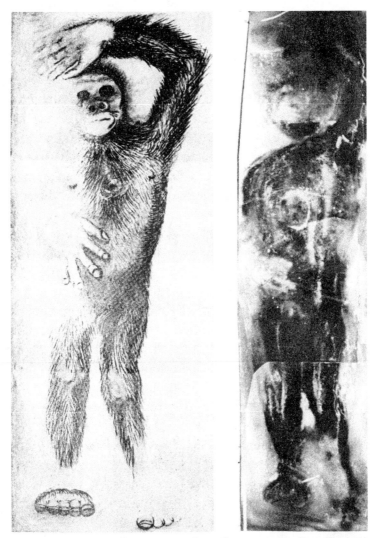

Ivan Sanderson's sketch of the pongoid man, based on a vertically referred and somewhat debatable method of measurement, shows some distortions, particularly in the exaggerated size of the hands and feet. The anatomical details, reproduced from memory, put excessive emphasis on the more simian characteristics (left). Putting together four partial photos of the specimen, taken from exactly the same distance, Heuvelmans obtained a composite image that enabled him to make much more precise measurements (right).

On this matter, my host, a geology graduate, was extremely helpful in confirming the artificial nature of the ice and in proving that the corpse could not have been frozen into a block of ice at sea but had actually been frozen in a freezer.

Having edited my note to make it completely unassailable, I sent it on January 14,

1969, to Prof. André Capart, the director of the Institut Royal des Sciences Naturelles de Belgique. Concerned with the ongoing slow decomposition of the specimen, I insisted on the urgency of the publication of my note.

After submitting my note to the scrutiny of Dr. F. Twiesselmann, head of the anthropological section of the Institute, Professor Capart wrote to me about the "great interest" my note had created. He characterized my note as "a model of critical analysis of a scientific problem" and accepted it with enthusiasm and told me it was already *in press*, given the exceptional circumstances. The Institute set a remarkable record in the annals of scientific publications by publishing my note within less than a month! Who could possibly have expected such speed?

As for me, I wanted to be absolutely sure that there was enough time for the American powers-that-be to take the steps required to have Hansen agree to yield his specimen before he found a way to make it disappear. I figured that three months would be amply sufficient. However it was important to keep the affair a secret; Hansen had to be taken by surprise. So, I obtained a promise from the Institute that news of my discovery wouldn't be released to the media before March 10, at the earliest. This would also give Ivan the opportunity of being the first, as was fair, to publish an article on our extraordinary discovery in a popular magazine.

While I was busy informing scientific circles in the hope that they would influence the administration, Ivan kept in touch with Hansen by phone to try to get his permission to have an x-ray specialist take some radiographs of the frozen corpse. At the same time, he also hoped to get Hansen's permission to publish an article, using his own photographs, in *Argosy*, where he was the Science Editor and which had financed his travel to Minnesota.

On January 16, Hansen came in person to visit Sanderson on his farm in New Jersey. It is there, on his own turf, where Sanderson managed to draw out from the ex-pilot some valuable information. As he was expressing to his visitor his surprise at having heard two different versions of the origin of the specimen—that of the Soviet trawler and that of the Japanese whaler—Hansen confessed that neither was true. He admitted, in fact, that he knew nothing about the original provenance of his hairy man. He added that he had first heard about the specimen from a bunch of old fly-buddies during a crossing of the Pacific to the Far East. Following their directions, he had gone to Hong Kong and visited a Chinese merchant with a British passport, an import-export specialist. There, in a vast commercial cold room, he had been able to see the specimen in an enormous bag of heavy plastic. He had asked for the price and made a counter offer, although he did not have sufficient funds. After returning to the United States, he had finally found someone who had advanced the funds necessary for the purchase. He had then returned to Hong Kong and brought the specimen back to the States. However, Hansen never managed to find out from the salesman anything specific about the origin of the specimen.[18]

Hansen had not come to provide Sanderson with more information, however. On the contrary, it was to inform him of the categorical refusal of the true owner to have any publicity or analysis of the specimen. No x-rays, no articles!

Deeply disappointed, Ivan then played his trump card: he told Hansen that I was about to publish a report of my examination of the specimen in a European scientific journal. Hansen was dismayed and furious. However, Ivan had scored a personal victory; now that the cat was out of the bag (in the words of journalist Tom Hall in his account of the events) Hansen no longer had cause to insist on Sanderson's discretion and no longer objected to his publication of his own comments on the enigma of the frozen man.

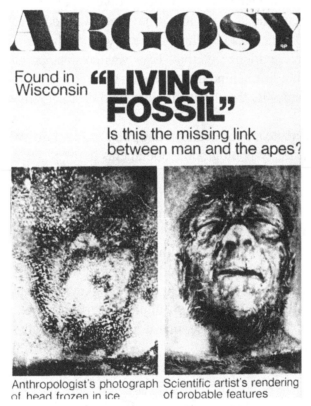

Ivan Sanderson's popular article in Argosy brought detailed information about the Iceman to the attention of the American public.

---

18   Later, on February 20th, Sanderson asked Hansen on the phone about the particulars of the Chinese businessman's business. The ex-pilot answered rather evasively: "Oh, he's an exporter.... he trades in just about everything, you understand, from marijuana to harder stuff, more or less...." That's all Ivan could learn from him. Anyway, Hansen absolutely refused to explain how the specimen had been brought into the United States.

Unfortunately, Sanderson, although science editor of *Argosy*, only had his article published in the May issue of that magazine, which would not come out until mid-April. Why this absurd delay that would deprive Sanderson of the glory of a scoop: the worldwide premiere of a sensational discovery? Simply because he wanted to first publish, in the April issue of *Argosy*, another article which would later no longer be of any interest.[19]

So it was that thanks to Professor Capart's extraordinary diligence, the article authored by Ivan Sanderson, the man who was at the center of our discovery, was to appear two months after my own note and a month after the world media took hold of it. My partner, Sanderson, who was an extremely proud man, took it as a personal affront, an injury to his reputation. This certainly didn't improve our relationship, which was already somewhat strained from our divergent views on how to proceed in this affair. By my haste he accused me of having thwarted his efforts to procure the specimen. All I had done was to adhere quite strictly to the plan that we had agreed upon; his notes were evidence of that. Ivan was so disappointed of having been, through his own fault, left behind and relegated to a secondary role in this affair that one might well wonder if, from that time on, he didn't actually work to hinder our efforts.

Sanderson had made a poor decision in opting to act alone, which kept him from bringing to bear the many photographic documents in my possession. A more careful study of those photographs would have saved him a number of false steps.

First of all, he wouldn't have mistaken the frosty oval surrounding the right hand for some kind of heel, as appears on the forelimbs of baboons, apes which have become more ground- than tree- dwellers. Then, he wouldn't have asserted that the hair was of the "agouti" type, with bands of light and dark pigments—a pattern unknown among primates—an appearance created by parallel strings of bubbles within the ice. These two details, noticed by an expert in monkeys and apes, led a number of zoologists astray by implying that the specimen was a quadruped, thus excluding any hominid, and even, and much worse, that it had been artificially implanted with the hair of another animal.

But that's not all. During our joint examination of the corpse, Ivan, who was an

---

19   This particular article was about an observation of a hairy biped in Wisconsin, and the subsequent discovery in the same region of footprints attributed to Bigfoot. On our drive back from Minnesota, we had looked into this affair in Fremont. I had been the only one to examine and photograph, at night, the footprints left in the snow in the middle of the forest, and I had been able to verify that they were a poorly crafted hoax. The prints had obviously been made by a joker wearing crudely-shaped frogman flippers. However, Ivan had been impressed by the report of the eyewitnesses and considered this a good topic for an article. Of course, there would be no point in publishing this article after the one on the frozen man, since the recurring stories of footsteps in the snow would then appear rather boring and of little interest.

excellent draftsman, had not made any sketches, probably relying on the photos that I was taking. He had only made measurements from above of various reference points. I have checked for myself the ineffectiveness of this method. The slightest movement of the head can lead to a tilt in the aiming axis of 5 to 10 degrees. With the eye 20 cm above the upper pane, there results at a depth of 40 cm below, at the level of the middle of the corpse, a shift of 4 to 7 cm, which is enormous.

This explains the important distortions in the drawings that Sanderson published first in *Argosy* and then in *Genus*. In the sketch drawn from the measurements he took, the outline of the body does not connect in some places (like the left leg) and the anatomical structure verges on the absurd (the left arm is twice as long as the right). As to the general appearance of the specimen, drawn entirely from memory many days after the viewing, it only vaguely resembles the original; what is seen is a creature with the head of a chimpanzee, with an overly slender body, too-long arms, and exaggeratedly large hands.

My own drawing was the result of a tracing of my photographs supplemented by an interpretation of the poorly visible areas. When it was published, Ivan, rather than admitting the understandable inaccuracy of his own sketches, tried to downplay the value of the photographic documents, claiming that they were dark and distorted and of generally poor quality.

Much worse, it is on the basis of the drawings sent by Sanderson on Feb 3 that two renowned specialists based their first diagnostic of the nature of the specimen. Dr. W.C. Osman-Hill declared that it seemed "more pongid than hominid," while Dr. John Napier said that "it belonged neither to one nor to another family," and that he preferred to "create a new family for it rather than try to force it into one of the older ones." These views from eminent primatologists, unfortunately misled about what was undoubtedly a man (whose proportions diverged only little from those of a classical Neanderthal), created an unfavorable impression among the international science community.

But I am jumping ahead. At that stage, Ivan and I were still working on our respective manuscripts. After translating my article into English, I sent a copy, respectively on the 5th and 9th of February to two of the world's most prominent specialists in primate anatomy, both old acquaintances, both British, and both working at that time, as luck would have it, in the United States. They were the two specialists mentioned above, Dr. William C. Osman-Hill, whom I had visited shortly before at the Yerkes Regional Primate Centre in Augusta, Georgia, and Dr. John R. Napier, who at that time led the Primate Biology Program at the Smithsonian Institution in Washington, D.C.

I could hardly have found anywhere in the whole world more competent researchers in primatology. My friend Osman-Hill had been working for about twenty years on the most monumental monograph on the comparative anatomy and taxonomy of primates: seven volumes had already been published since 1953. As to John Napier, with whom I had participated earlier in a BBC program on the Snowman, he was not only the author of an excellent textbook on living primates, but also an expert on the extremities of hominids.

Thanks to the support of these experts, the Smithsonian Institution soon took a strong interest in our discovery.

John Napier had immediately contacted Sidney Galler, the scientific secretary of the Smithsonian. Galler was very open-minded and understanding and was immediately aware of the importance of the specimen, which was such that it should be considered as part of the cultural heritage of humanity. That would justify the use by the American government of its right of preemption for the purchase or even confiscation of the specimen. That's probably why Sid Geller first asked Napier if the FBI had been made aware of the affair. Having been assured that Sanderson had contacted the local federal police at Morristown (New Jersey) on January 18, he felt that the Smithsonian could legitimately request the assistance of the head of the FBI, J. Edgar Hoover.

At the Smithsonian, a whole team of specialists was assembled under the leadership of John Napier to perform an autopsy and a detailed examination of the hairy man. However, they first had to get their hands on it.

Alas! Alerted by Ivan's imprudence, Hansen had already heard about the forthcoming publication of my report. He had also found out, because Ivan had read long excerpts of my note to him over the telephone, that I proved not only that the hairy creature was human, but that it had been shot and artificially frozen.

Hansen did not take long to react. On February 20, he called Ivan to tell him that the "owner," enraged by the developments, had spent ten days with him and ended up taking the hairy corpse away in a refrigerated van.

Before leaving, however, he had left behind a replica created earlier at great cost and difficulty "just in case he had problems some day." Hansen added: "At the time, I thought this was going a little far, but I understand what he meant."

Hansen was soon to repeat all that to the media, which was now taking a keen interest.

Up to March 10, my science report, published a month earlier, had been available only to scientists, whom I hoped to convince of the need to alert administrative authorities. However, on the agreed upon date, my note was released to newspapers by the Institut Royal des Sciences de Belgique, and as early as the following day, our discovery was front-page news in the Belgian press. It was soon to trigger a passionate discussion in the British press and subsequently all over the world. All of which was of great concern to Hansen, who always seemed to be up to date on the news in the backwoods of Minnesota.

So it was that on March 13, the general secretary of the Smithsonian Institution,

S. Dillon Ripley, wrote a personal letter to Hansen, asking him to allow Smithsonian scientists to examine the famous specimen, which "according to the description and photographs of Dr. Heuvelmans" seemed to be "of great interest to the scientific community." Ripley vainly explained to him how much he would appreciate his collaboration in this study, which was likely to "turn out to be a major contribution to human knowledge."

Hansen answered that the specimen was now in the hands of its owner who "wished, for a variety of reasons, to remain anonymous." Hansen also thought that there was little hope to convince him to change his mind and to allow "any kind of scientific investigation." He added that they had created together an "illusion show" for 1969 that would "look very much like the specimen photographed by Dr. Heuvelmans," but he couldn't be sure if "the original exhibit would ever again be shown to the public."

According to the people who met him at that time, Hansen gave the impression of a man cornered. Following a complaint that he might be hiding a human corpse, the local sheriff had visited him. However, since American law does not allow a search warrant to be issued unless there is sufficient prior evidence that some crime has been committed, all Hansen had to do to avoid a search was to declare that he only had in his possession a dead monkey, or some mannequin. He actually had no trouble convincing the local sheriff that this was the case. But soon afterwards, it was the FBI who came knocking and asking questions, which didn't seem at all related to this business. There was no doubt that he was under investigation. The reason why he was so terrified was, of course, because of all the illegal connections apparently related to his exhibit. But exactly just which ones?

Just for the sake of completeness, let's hark back to the possibility of a false corpse. As I mentioned earlier, Hansen was quite willing to admit that the specimen might be artificial and, rather curiously, seemed to favor that possibility. To exhibit something for money under false pretenses is clearly a fraud and hence punishable by law. But who would bother suing the exhibitor for having presented his specimen as a "medieval man" or "preserved for centuries"? The law relative to fraud and misrepresentation is difficult to apply strictly to side shows and carnivals (and even to advertising in general), and many exhibitors of strange phenomena would have to pack up if it were scrupulously applied. It is indeed rare that a specimen presented as the "most," or the "real," or the "only" this-or-that is exactly what it is claimed to be.

The sudden disappearance of the specimen and Hansen's panic were most likely to have more serious causes.

First, it was clear that a human being, in the widest sense of the term, had been shot, as its wounds demonstrated, and subjected to some violence. Nevertheless, in this case it would be unlikely that a charge of murder could be made to stick: hunting accidents occur frequently and are rarely penalized.

Given its appearance, everyone would understand that the hairy man could have been shot by mistake, having been taken for a bear or some anthropoid ape; a dangerous creature. One might even legitimately claim self-defense.

What was rather more difficult to justify than the death of the hairy man was the fact that it had been frozen so as to be shown in county fairs. Possession of a human corpse must be immediately declared to the law and can only continue for precise reasons, and with the approval of State and Federal authorities. Transporting a corpse across state lines without appropriate documentation is a federal crime.

The mystery surrounding the precise origin of the specimen also suggested that one might be trying to hide some illegal activity. Instead of having been purchased in Hong Kong, as Hansen claimed, might the hairy man have been brought in from China? At the time, the US had not formally recognized that country and there was an embargo on all goods coming from it. All trade with Mao's country was forbidden for American citizens under threat of harsh penalties.

That was, of course, only a supposition. One fact however was to be revealed upon inquiry: *there was no mention at all in the records of US Customs of the introduction of the hairy man into the United States*, whether dead or alive, within a block of ice or inside a cage, wrapped up like a sausage, or asleep from drugs and injections.

It is forbidden to bring anything into the United States without declaring it, even when there is no custom duty to be paid; smuggling is a federal crime. That also applies if the hairy man is considered a person rather than an object. It is illegal to enter the country without a visa and to stay without authorization. To assist in bringing a person illegally into the States is also a federal crime.

The situation is made even more delicate if we imagine that the specimen could have been introduced fraudulently into the country onboard a United States Air Force flight, which was entirely possible for an experienced fighter pilot. To use military facilities to commit a crime is a matter for a court-martial! That is absolutely not the thing to do for a career soldier, especially if he is soon to retire and collect a comfortable pension.

In theory, one could also wonder if perhaps originally the specimen might not have been stolen abroad. That might explain why the owners were willing to put up with publicity in local publications but were wary of publicity in wide circulation media that might attract attention in other countries. Had that been the case, they would have found themselves in the embarrassing situation of art collectors in possession of dearly bought items stolen from some famous museum and on Interpol's wanted list. Such people can only enjoy their treasures privately or in the company of trusted or ill-informed friends.

Perhaps the key to the mystery consisted in a combination of many of the above hypotheses, even of all of them, which would only make it more difficult for the owners. Just for fun, let us imagine the worst situation. Having been captured in

Asia, the wild man was purchased in Communist China, then stolen from its rightful owner and subsequently assassinated when trying to escape. Frozen to prevent decomposition, fraudulently brought to the United States as cargo on a military flight, placed in a block of ice and exhibited in country fairs, illegally transported across state lines, and kept without a permit…that would be enough for someone to end up with hundreds of years of accumulated penalties!

I was hoping that Hansen might have better explanations to justify his silences, his lies, and his bizarre behavior. I would have been truly sorry to see any serious inconvenience befall a man who had kindly welcomed us. But I would have been sorrier and even seriously incensed if, through his fault and that of the owner, the specimen were to rot away or disappear forever like an embarrassing witness. There was no reason whatsoever that could excuse hiding such an important item from the scrutiny of science and the enlightenment of all of humankind.

Chapter 5

MORE COMPLICATIONS

Lightning interventions by the FBI and the Smithsonian

"Here we go round the mulberry bush…"

Faced with the attitude of a carny who refused to grant researchers permission to examine the "ape-man" in the news in the spring of 1969, the Smithsonian Institution officially asked J. Edgar Hoover, the head of the FBI, for help in locating the missing specimen and, if necessary, to seize it. In addition, Dr. John Napier, speaking as a representative of the famous scientific institution, informed the American media of the affair. From now on, Hansen was to be visited from time to time by nosy policemen and submitted to the constant harassment of the press and curious onlookers.

Just before conveniently disappearing for three weeks, supposedly on holidays, Hansen was called on the phone on March 25 by Marlise Simons, of the *Times* of London. On the defensive against the various suspicions raised against him, he offered a range of confusing explanations. He repeated that "because of the unfavorable publicity, from now on only a model will be shown to the public rather than the real creature" when the exhibit would be shown again next summer. (A model of what? people might ask.) Hansen continued: "We do not have a permit. The law forbids transportation or exportation of a corpse, but there is no law against transporting an ape. If it was truly of flesh and bone, who could say today whether it was a man or an animal? Of course, the scientists will say it was a man, just so as to be able to take it away from us, but the owner has excellent lawyers and I believe he will take the case to court."

What did Hansen mean by these obscure declarations? That his specimen was just an ape? Or that it was not, or no longer was, of flesh and bone? What would turn out to be clear was that the mysterious Californian owner truly did indeed have some excellent lawyers who would know how to take advantage of those articles of the American code of law protecting the rights and the privacy of citizens.

To refute the charge of smuggling, it was enough for Hansen to insist on proof that the specimen had been imported in the United States, and not captured, purchased, or even fabricated within the country. Regarding the charge of homicide, he could insist on proof that it was a man and not an ape that had been killed, and even proof that the corpse was not just a fabrication.

While such proofs were obvious to Ivan and I who had examined the specimen, they were not so for the police who, once the specimen had vanished, could think that our diagnostic was partly based on opinion rather than fact. Furthermore, the fear of ridicule held them back: What would the FBI look like if it ended up with nothing more than a rubber or plastic model?

This is why, around mid-April, the FBI informed the Smithsonian that it could not help because the case was beyond its jurisdiction. It couldn't act until there was some proof that a crime had actually been committed. Mere suspicion was not enough to justify the FBI's intervention.

Emboldened by this news item, which was widely reported in the press, Hansen suddenly resurfaced. Back on his farm in Rollingstone, he gave a telephone interview to a television station. On April 20, he went as far as inviting the press to his home. Wearing his usual Texan hat and hiding his eyes behind sunglasses, he formally presented to the journalists, including Gordon Yeager of the *Rochester Post-Bulletin*, the frozen creature currently in his trailer. He emphasized, however, that this was not the original being, but "*a man-made artifact*." He gave ambiguous answers to probing questions. He allowed the photographers to take as many pictures as they wanted.

Really, Hansen was only carrying out what he had announced to Sanderson two months earlier and repeated to the interviewer from the *Times* of London. Providential incidents—perhaps rather too much so—were soon to reinforce his allegations.

At the beginning of May, George Berklacy, public-relations man for the Smithsonian, received a phone call from the director of a wax museum in California. That gentleman told him that one of his employees, whose name he could unfortunately not reveal, had collaborated in April 1967 in the fabrication of an ape-man in foam rubber for Frank D. Hansen. It was his employee who had inserted the hair of a bear into it.

This "spontaneous" revelation was perfectly timed to confirm Hansen's latest declarations. Not that there was much cause for such a fuss; it merely confirmed what was already known, namely that a copy of the specimen had been fabricated earlier so that it could replace the original "in case there were to be problems."

When Napier called Sanderson about it, the latter merely shrugged and declared that the wax museum's call was clearly a ploy aiming at discouraging the Smithsonian's interest in the matter. Sanderson also claimed that he had found another professional model-maker who had crafted a similar model, again for Hansen, but in April 1969. In any case, the specimen that he and his friend Bernard had carefully examined was certainly not made out of rubber because it was rotting away with a noxious smell. Furthermore, added Sanderson, even if that specimen had been a fake, which it couldn't be for a variety of other reasons, it would necessarily have been constructed from parts taken from living beings. And he went on to explain, in that tone of his that bore no contradiction, how he would have proceeded to make such a fake using the hide of a very clear-skinned chimpanzee, spread over a human skeleton after some modifications to the hands and feet using a glove-spreader.

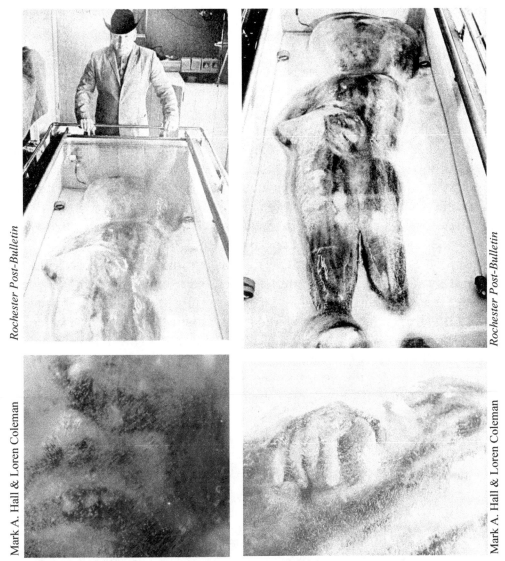

Rochester Post-Bulletin

Rochester Post-Bulletin

Mark A. Hall & Loren Coleman

Mark A. Hall & Loren Coleman

*Terrified by the publication of Heuvelman's report, which called upon him to explain
the origin of the specimen, Hansen immediately withdrew it from view.
Two months later, he would present it again for public scrutiny, but now claiming
that it was only a copy of the original. Actually, he had only thawed it and allowed it to
freeze again, which affected its appearance somewhat. The ice was clearer and
the big toes more spread apart than formerly. The mouth was now open,
revealing the teeth. And the right hand was now more visible.*

It was also to demonstrate his open-mindedness in evaluating the nature of the
original specimen that, while affirming its irrefutable authenticity, Ivan nevertheless
claimed to be able to fabricate one just like it. He was also laying it on thick, for
as I mentioned earlier, this was impossible because there are no large pale-skinned

chimpanzees. Ivan's boast was to cost us. Claiming such ability in the presence of the Smithsonian's representatives was like putting on the noose with which he would hang himself. Of course, all disbelievers would remember his claim and declare that: "Since Sanderson, one of the discoverers of the specimen, admits that it would be quite possible to make one just like it, it's obviously a fake."

Incredible as it may seem, even John Napier was taken in by the obvious maneuver of the "spontaneous" phone call to the Smithsonian and by Sanderson's arrogant claim as to the technical feasibility of fabricating a convincing likeness. Napier thus gradually came to the conclusion that there had never been more than a single frozen man, namely an extremely skillfully made model. As he had never examined the specimen, this was for him a possibility worth considering. Besides, due to circumstances completely extraneous to this business, it suddenly appeared advisable to distance the Smithsonian from the search for the hairy man.

It happened that the prestigious institution had on numerous occasions been strongly attacked in the press and on television over some of its activities. It was criticized for having participated in experiments financed by the Department of Defense on the migration of some birds in the South Pacific; in case of conflict, these birds could be used as vectors of disease against the enemy. Public opinion was currently hostile to the Smithsonian, and it was crucial that the reputed institution should not be ridiculed in case the specimen turned out to be a rubber dummy peppered with the hair of a bear or a chimpanzee and modified by plastic surgery.

Thinking that he had left the situation with the Iceman in good hands since the Smithsonian had taken charge of the affair and the FBI was inquiring as to its origin, Heuvelmans left the United States somewhat lighthearted, to embed himself, as he had planned, into the jungles of Guatemala. It's only when he returned to France that he learned that American officials had dropped the ball. (Photo courtesy Rochester Post-Bulletin)

As he wrote to me later, John Napier would have liked to be able to explain to me the reasons behind his change of heart, but I was far away. Fully confident in the effectiveness of the Smithsonian as an official institution fully capable of securing the specimen, I had left Washington unconcerned on March 27, crossing Mexico in a Land Rover with my friend Jack Ullrich, and I was now deep in the Guatemalan jungle. Unable to reach me soon enough—or perhaps happy that he couldn't—the man whom I had trusted with my scientific interests did not hesitate, by May 8, to put out a press release, which, for the benefit of historians of science, is worth quoting in its entirety:

> The Smithsonian Institution is no longer interested in what has been called the Minnesota Iceman[20] because it is convinced that this creature is nothing but a fairground fabrication made of foam rubber and hair. A reliable source, which the Smithsonian is not authorized to reveal, has provided information regarding the owner of the model as well as to the date and place of its fabrication. This information, together with some recent suggestions by Ivan T. Sanderson, the scientific writer and original discoverer of the iceman, on how to construct such a creature have convinced us beyond reasonable doubt that the "original" model and the alleged current "substitute" are one and the same.
>
> The director of the Primate Biology Program of the Smithsonian, Dr. John Napier, points out that this institution has always maintained an attitude of skepticism combined with an open mind, and that its only interest in this affair is to discover the truth, which it is reasonably certain is as stated above.

I may be a complete idiot, but I don't understand how one can come to the conclusion that a specimen is a fake by *combining* the fact that there exists a rubber model with the fact that it is only possible to make one by stretching the skin of a chimpanzee over a human skeleton. One should chose between these two contradictory possibilities: a blown-up dummy studded with monkey hair, or a modified ape. By avoiding the choice, the agents of the Smithsonian betrayed either their bad faith or their blindness. As I'm easy-going, I'll grant them the choice between dishonesty and stupidity.

---

20   *Iceman* may mean "ice salesman" as well as "man of ice." Similarly, "snowman" may mean "man-made out of snow" or "man of snow." Right from the beginning, and in imitation of the Abominable Snowman, a ridiculous moniker was used for the specimen, perhaps with the more or less conscious hope that it wouldn't be taken seriously. I couldn't tell for sure who coined the name, but I saw it used for the first time in a letter to me from John Napier dated February 14, 1969, before any mention of it in the press.

Was it a sudden impulse of decency or of scientific probity that led John Napier, immediately after the publication of the Smithsonian's press release, to inform some journalists that, personally, he remained keenly interested in the frozen man and eager to examine it. He added: "It is difficult to believe that Dr. Heuvelmans would have been so easily fooled."

Further repercussions were to show that good old Heuvelmans had not been wrong at all. However, Napier's unfortunate move had caused irreparable damage. Although somewhat reserved in its original statement, the Smithsonian's press release was to be repeated peremptorily in the press. Some went as far as writing that an autopsy of the specimen that Ivan and I had described had already been performed to reveal that it was merely a rubber dummy. A myth gradually took form to the effect that the famous "frozen man" was just a hoax. Needless to say, this revelation was often accompanied by unflattering, mocking and even insulting comments about its discoverers. It was in vain that later developments showed that we were right. Public opinion was now firmly against us.

Let's go back a few days. On May 5, Hansen started again to exhibit to the public a hairy frozen creature, first in a commercial center in St. Paul, Minnesota. His trailer bore a new sign, even more naïve than the previous one:

SIBERSKOYA CREATURE (a Siberian creature)
A MANUFACTURED ILLUSION, AS INVESTIGATED BY THE F.B.I.

In spite of that advertising, which lent the exhibition a rather frivolous aura, Professor Murrill, of the Anthropology Department of the University of Minnesota, hastened to go and examine the controversial specimen. He was so impressed by what he saw—as both Ivan and I had been—that he offered Hansen a rather large sum of money, quickly declined, to acquire the specimen on behalf of the university. Not that he was 100% convinced of the authenticity of the item (he even found that the hair had a *rather suspicious appearance), but he had been seriously shaken.* In any case, he had done all he could to try to salvage what, in his mind, had a good chance of being an extremely important specimen. This fundamentally scientific attitude was in sharp contrast with the incredulity, lack of interest, wait-and-see attitude, or complete dismissal of so many others.

"It's the damnedest thing I've ever seen!" admitted Professor Murrill to Dr. Napier. However, this new testimonial, far from tipping the latter's doubts in favor of the authenticity of the specimen, tended even more to confirm his views that he was in the presence of a single "fantastically successful" model.

Later on, Hansen moved his exhibit to Grand Rapids, Michigan, where a crew from *Time-Life* publications came to film it. After having seen that film, John Napier noted that the specimen on exhibit differed in a number of points from the one I had studied and photographed. For example, the mouth was now open and showed a full row of upper teeth, and the big toe was further apart from the others than before.

Having finally returned to France from Guatemala, I wrote to Napier on May 29 to protest his disastrous initiative, which I saw as "a desertion before the assault," a rout provoked by the first sign of a reply from our detractors.

As a vindication, as much for the eyes of S. Dillon Ripley, general secretary of the Smithsonian, as for Ivan and myself, Napier sent each one of us a memorandum in which he explained all the elements on which he based his doubts. Nevertheless, he recognized that his reasoning *might* be erroneous. He even wished, he claimed, that it were.

From New Jersey, Sanderson sent to Napier on June 19 a counter-memorandum in which he noted a number of points that had simply not been taken into account, facts that had never been verified or controlled, which were taken as solidly established but were merely based on presumptions.

Almost at the same time, within a day, I sent from France my own counter-memorandum to Napier. I listed all the facts that remained unexplained and were even *unexplainable* from the perspective of a hoax.

Later on, after Ivan and I had exchanged copies of our respective responses, we were struck by the nearly perfect match of our arguments, a clear sign of their relevance.

There was only one point on which my views diverged from Sanderson's, as well as from all others who had looked into the matter, and that was on the nature of the specimen exhibited by Hansen after April 20. I was the only one to believe that *it was still the actual corpse*. True, I had a definite advantage over everyone else—I was the only one to have many excellent photos of the original exhibit.

I had been sent a few color slides of Hansen's new exhibit. After a careful comparison with my own, I had to agree with the evidence: *it was the same and only specimen.*

In its fine structure, the coating of ice that covered the hairy man appeared completely new. It was noticeably more transparent than before in some areas, especially around the forearm, where former traces of frost had completely disappeared; the frosty circles surrounding the left hand and in the middle of the chest had also vanished. The specimen had been thawed and before freezing it again, this time with distilled or boiled water to eliminate bubbles, his mouth had been opened enough to view the teeth and the big toes had been spread out a little more.

Its complexion had become more ashen, which is easy to understand; decomposition proceeds extremely rapidly when the temperature rises after a long period of freezing. A lot of blood had been smeared over the face in the process, apparently coming out of the eye sockets, the nose and mainly the mouth.

How could I tell it was the same specimen? Well, it was as much through how it differed from the original than by how it looked like it.

Here's my argument. If, for example, you leave a room where you have been with a familiar person and return a few hours later, how can you be sure that it is the same person that you are seeing? By the perfect identity of appearance with what she was like when you left? Surely not. That person certainly moved, changed position and expression. She might have moved to another room, or perhaps changed clothing. She might have also changed hairstyle, perhaps even dyed her hair. You might find that surprising, but you would not think that you are meeting someone else. So, this person looks different, perhaps even very different, but she is certainly the same person.

How can you tell? First of all by her general physical appearance, which consists of a number of movements, like how she holds her head, or her arms, and also the shape of the mouth, nose, eyelids, and all the wrinkles of the face. Actually, the only traits that remain perfectly identical in that person are the small, often nearly insignificant, details such as the setting of the eyebrows, the arrangement of beauty spots, some wrinkles, a mole, or a scar...

Truly, if after having left the company of someone for a few hours you were to find her absolutely identical, in the same position she was in, with the same expression, as if frozen in time, you would have reason to be concerned. You could then have well asked if she has not perhaps been substituted by a wax figure.[21]

This was exactly the case with Hansen's new exhibit. I could see that the specimen was truly the one I had photographed from a few minor details, apparently insignificant, and which one would not have thought of duplicating in a model: imperfections in the skin, such as scratches or pigmented spots; the relative arrangement of eyelashes or eyebrows. All this was also evident through the various modifications made to the specimen without really changing its identity.

So pongoid man had been thawed to change it slightly to convince Ivan and myself, as well as everyone who had seen my photos, that this was a different specimen, namely a replica. The big toes had probably been spread out so as give it a more simian appearance and avoid the suspicion of murder.

If that was indeed the case, why did everyone believe it was a fake? Simply because Hansen had said so.

If the Mona Lisa, the real original painting, were to be for sale at the flea market, and presented as being authentic, people wouldn't even bother looking at it. On the other hand, if it were presented as a copy, no one would doubt it and no one would

---

21   Take another example. Whenever a policeman asks a suspect or a witness to repeat his statement many times over, it is as much to discover variations than to make sure the basic information remains constant. Should the interrogated person repeat each time the same story word for word, the policeman would know that the testimony has been memorized and that it is likely to be incorrect.

insist on an expert appraisal. Furthermore, if it was then offered at a ridiculous price, although much too high for a mere copy, no one would buy it, and again no one would bother with an appraisal.

People only want to believe what is normal, what reassures them.

When, on the strength of my professional experience and my university degrees, I had affirmed that the specimen, which I had examined, was an unknown hominid, many people had loudly requested x-rays or an autopsy to confirm my diagnostic. Fair enough. However, when a mere carny had claimed that what he was exhibiting was a made-up model, everyone agreed. A double standard indeed, which shows that it's not always the weightier evidence that tilts the scales.

Simple logic had led me to suspect the truth long ago. From the very beginning of the affair, I had been struck by its cleverness, which tickled my fancy for detective stories. I had even written:

> Rather than trying to hide or to dispose of a bothersome corpse— never an amusing and often a disgusting task—why not simply exhibit it to the public with much publicity and hoopla?
>
> What an admirable application of the principle expounded by Dupin in Poe's "The Purloined Letter": one cannot see what is hiding in plain sight. What genius in the art of crime! What a novelty for a hoaxer to present the real item as a false!

The same cunning must have worked its magic in hiding the original specimen. While the investigators were looking all over the United States to find it, I thought of "The Purloined Letter." What was the surest place to hide the hairy man, a place where no one would dream of looking? Of course, where it was clearly visible to everyone, *in the glass-covered coffin where Hansen was exhibiting his manufactured replica*!

What this meant is that the so-called replica wasn't one: it was the real corpse.

But then why the hell have a rubber model made, or perhaps even two?

In my view, no rubber replica had ever been exhibited publicly for the simple reason that it would not have been convincing. By its very nature, a replica would be too imperfect to fool anyone with even only a little expertise. Whatever the degree of perfection attainable in a rubber model studded with real hair, it cannot pass a rigorous examination. As I said earlier, it is practically impossible to imitate real skin, with all its imperfections and irregularities.

In spite of that, a manufactured replica was an ideal guarantee of security. Should the demand for a careful scrutiny become irresistible, one could always bring out the false specimen. Even better, as a means of thwarting the efforts of investigators, it would suffice to show the receipts documenting its fabrication or show photos of its successive stages. (That's obviously what Hansen had done to discourage the sheriff who came over to enquire about the "human corpse" he was thought to have in his possession.) It was also possible, if and when necessary, to produce statements by various artisans who had contributed to the fabrication. (This is what happened, out of the blue, when the Smithsonian took too close an interest in Hansen's specimen.)

As to the second model (I doubt that there really was one), it could have been thought to be useful following the modifications brought to the original. It would then have been made to conform to this new appearance. With two rubber fakes in hand, one would have been able to cover all angles and never have to produce the flesh and bone specimen. So, one could have brought out a (false) original specimen, allegedly purchased in good faith as a complete mystery, and an acknowledged replica, slightly different, and fabricated to protect from the possessive owner from public scrutiny. There would then have been a false "real specimen," a real "false specimen," and a false "false," which was of course the real "real." Go figure!

My arguments as well as my conclusion were sent to John Napier in a counter-memorandum on June 20th, with a request to forward a copy to S. Dillon Ripley.

What I was soon to learn from the US confirmed my suspicions.

At the beginning of the 1969 summer season, Hansen had crossed the Canadian border to show his "manufactured" illusion in a number of provincial fairs, including Calgary, Edmonton, and Winnipeg. Having completed his tour, he was preparing on a sunny Sunday afternoon to return to the States via a border crossing in North Dakota. But US Customs barred his way and refused to allow him to bring his trailer into the United States. If it was true, as it was believed, that he was carrying a "humanoid" creature, he needed permission from the US Surgeon General to transport it.

Hansen resisted and claimed, to no avail, that the specimen was only a fabricated model; he was told to prove it by allowing a small sample to be taken … a post-mortem biopsy. He refused, claiming that this would ruin his exhibit. In that case, he couldn't come in.

Finally, Hansen started calling everyone for help. He called, among others, Sanderson, with whom he had remained in excellent relations. Ivan, who believed like everyone else that Hansen was only carrying a replica, couldn't understand his panic. Since the exhibit was only a bran-filled rubber replica, as Hansen had claimed, just let the customs officers take an x-ray. The exhibit will not be damaged and the truth will out.

No way, Hansen shouted; he absolutely refused. The owner would never allow it, he says. (As if it was necessary to inform him of this inoffensive procedure, which would leave no trace.)

Finally, Hansen decides to call the owner, in California, so that the latter might ask his "excellent lawyers" to intervene; he even calls his Senator, Walter Mondale, in

Washington, D.C.[22] After twenty-four hours between border posts, Hansen is finally released and returns to the States with his precious cargo.

Over the previous months, perhaps to assuage my legitimate impatience and my growing anger at the apparent ineptitude of scientific and administrative powers, I had been told and retold that although the Smithsonian and the FBI had *officially* abandoned their interest in this affair, both institutions actually remained attentive and that others, like the US Customs Service or the Department of Public Health, were only waiting for an opportunity to seize the specimen and determine its actual nature. This incident proved that this was certainly not the case. This time, there was a perfect legal motive to submit Hansen's specimen to some scrutiny, even if only by x-rays, and the occasion had been missed because of political or some other pressures. It was now clear that Hansen had friends in very high places.

This time, I began to really lose patience. How was it possible for the American authorities not to realize that the rest of the world, unbiased scientists as well as the general public, would some day severely condemn them if they were to allow the destruction or the loss of what was undoubtedly the most precious anthropological specimen they ever had in their hands. I was despairing and elaborating the darkest thoughts about how to draw attention to this scandal.

One day, however, it seemed that the very owner of the specimen might have finally realized the seriousness of the situation. In September a faint hope appeared.

After being out of the picture for a few months, following the close call at the Canadian border, Hansen was again exhibiting his specimen in a variety of commercial fairs in the United States, notably in Minnesota. He also seemed to have regained his self-confidence. The frozen man was no longer presented as "a total mystery" or "a manufactured illusion." The advertising board clearly stated:

THE MISSING LINK,
THE ONLY "HOMO PONGOIDES" IN THE WORLD.
A SPECIMEN WHICH AMAZED MILLIONS OF PEOPLE
AND PUZZLED THE ENTIRE SCIENTIFIC WORLD

---

22   Walter Frederick Mondale, then 41 years old, had been as US senator since 1964 and was a lawyer by profession. He had been called to the bar in 1956 and had been in private practice until 1960 when he was appointed attorney general of Minnesota. We recall that the attorney general is the ranking justice official in his State and has, by that fact, powers exceeding even those of the governor. However, he could not intervene in another state nor in federal issues, which would be under the jurisdiction of the US Attorney General.

Mondale has a sterling reputation for integrity among Democrats; some even saw him as a future president of the USA. It is then not surprising to find him blacklisted by the White House.

This sudden boldness was soon to find an explanation. Hansen announced to the media that his three-year contract (which he had made a vague reference to when speaking to Ivan and I) was to expire the following spring. Winter months being particularly unproductive, he would end his exhibits in December 1969. He then revealed that the owner of the specimen had finally decided to submit it to a thorough scientific analysis, which would take place in a laboratory that he would set up, where a few anthropologists would be invited to assist him in this task.

In a word, Hansen was finally and publicly admitting that *there existed a real specimen of great interest to science.* One does not set up at great expense a research laboratory simply to dissect a rubber dummy, and one does not invite learned anatomists to examine the sawdust with which it is stuffed.

However, was there really cause for applause at Hansen's revelation and at the laudable intentions of the wealthy proprietor? I personally suspected that the grandiose plan to set up a special laboratory might only be a delaying tactic or some strategic withdrawal leading to substitution of some other object without much scientific interest. Come on now! Why a private laboratory? Were there not within the United States perfectly well equipped labs—the Smithsonian, for example—capable of examining the corpse of the hairy man under the best conditions?

After a whole year of "abracadabra," I had good reasons to be suspicious. Nothing had changed regarding the legal status of the specimen. If it were, as I professionally suspected, an anatomical specimen of extraordinary importance, Hansen and his patron would have to explain its origin and how it might have been surreptitiously introduced into the United States.

Unless…ah, yes! Unless, of course, they declared that *the pongoid man had never been fraudulently brought into the country…* It was sufficient to claim that it had *been shot inside the country* and thus a member of the local fauna.[23]

What was to happen was exactly what one could expect given all the facts and, of course, a bit of logical thinking.

---

23   It was precisely to leave Hansen an escape clause, and to allow him to yield his specimen to science without serious legal problems, that I had in my scientific note evoked a hypothesis, which I nevertheless considered incorrect, regarding the origin of the hairy man: "Nothing proves," had I written, "that it had not been killed elsewhere, even in the United States." My only interest in this affair was scientific, and if I had to take an interest in legal issues to resolve it, those issues were nevertheless not my business.

Chapter 6

HANSEN'S STORIES

A pseudo-confession of a murder launches a useless expedition

" ... a lie which is all a lie may be met and fought with outright.
But a lie which is in part a truth is a harder matter to fight."
— Alfred Tennyson, "The Grandmother"

After Hansen had told the press that the corpse of the pongoïd man was finally going to be submitted to a thorough scientific analysis, in a laboratory specially created by its owner for this very purpose, months went by without any news about it. And then suddenly, in July 1970, a sensational development! The American magazine *Saga* published a long article by Hansen with the dramatic title: "I Killed the Ape-Man Creature of Whiteface."

Nothing could be more edifying than to read this long confession—a small jewel of horror literature—which I can unfortunately not quote in its entirety. Here is what the article says, in essence, which I first quote without comments (those will come later), but with great doubts as to the authenticity of the alleged facts.

The crucial event, the murder of the ape-man, was said to have taken place in 1960, at the very beginning of the deer-hunting season, in the wooded and swampy region of Whiteface Reservoir. This lake is situated about 100 km (60 miles) north of Duluth, Minnesota, where Captain Hansen's squadron was posted. Northern Minnesota is rich in game, and many officers of the squadron—specifically, Major Lou Szrot, Captain Frank Hansen, and Lieutenants Roy Aafedt and Dave Allison—had organized a hunting party on the opening day of hunting season.

On that first day, no one had seen any game, but on the next day, at dawn, Hansen had wounded a large doe with his Mauser 8 mm gun,[24] and he set out alone on the trail of the wounded animal. After an hour of a fruitless pursuit, he was ready to give up and turn back when he heard a strange gurgling noise nearby. Thinking that he had heard the doe choking in her own blood, he walked towards the source of the noise: *"Suddenly I froze in horror!"*

> In the middle of a small clearing were three hairy creatures that
> at first looked like bears. Two of these creatures were on their knees,
> tearing at the insides of a freshly killed deer. The deer's innards were

---

24   The article in *Saga* speaks of "my customized 8 mm Mauser," meaning, in the slang of hunters or hired killers, a military weapon transformed by a gunsmith into a hunting rifle.

scattered around the clearing and the "things" were scooping blood from the stomach cavity into the palms of their human-like hands. Raising their cupped hands of fresh blood to their mouths, they swallowed the liquid.

The third creature was about 10 feet away, on the edge of the clearing crouched on its haunches. It was obvious that he was a male, of similar stature as a man. Absolute horror gripped every muscle of my body as I stared at this frightening tableau before me. I felt as if my body had turned to stone.

Without warning, the male leaped straight into the air from its crouched position. His arms jerked upward, high over his head, and he let out a weird screeching sound. Screeching and screaming, he charged toward me. I cannot remember aiming my rifle, nor do I recall pulling the trigger, but a bullet must have slammed into the beast's body.

As blood spurted from his face, the huge creature staggered, seemingly stunned by this unexpected happening. I do not recall ejecting my spent shell, nor do I recall firing my rifle again. In many sweat-drenched nightmares, however, I have vividly envisioned the blood-covered face lying on the ground beside the mutilated deer. I have absolutely no recollection of ever seeing the other two creatures again. They seemed to have vanished into "thin air."

Blind with fear, I started to run. I dashed over the swampy terrain not knowing or caring in which direction I ran. My only thought was to get away from those horrible "things." I stumbled, fell, picked myself up, and fell again. I thought they were right behind me. Finally, I fell onto the frozen marshland completely exhausted, not caring if the creatures caught me. I lay there waiting for the attack.

I have no recollection of time and perhaps my mind blanked out. When I regained composure there was only the natural silence of the swampland. I wondered if I hadn't fallen asleep and dreamed the whole thing...

It was only around noon, after meeting with other hunters who helped him find his camp, that Hansen caught up with his mates. Of course, they laughed at him for getting lost. To explain, he thought of relating his adventure, but did not dare. He was afraid to be thought of as mentally disturbed, that he might be forbidden to fly, or that he might be dismissed from the Air Force. After all, he had only five years to go before drawing the generous pension he felt he deserved after twenty years in the USAF.

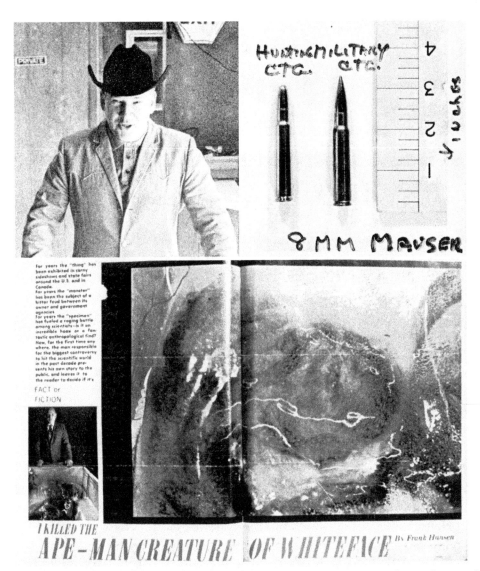

HUNTING MILITARY CTG. CTG.

8 MM MAUSER

For years the "thing" has been exhibited in carny sideshows and state fairs around the U.S. and in Canada.

For years the "monster" has been the subject of a bitter feud between its owner and government agencies.

For years the "specimen" has fueled a raging battle among scientists—is it an incredible hoax or a fantastic anthropological find? Now, for the first time anywhere, the man responsible for the biggest controversy to hit the scientific world in the past decade presents his own story to the public, and leaves it to the reader to decide if it's

FACT or FICTION

I KILLED THE APE–MAN CREATURE OF WHITEFACE By Frank Hansen

After having nearly given permission, at a boundary crossing, for a thorough examination of a specimen that he claimed to be a fake, Hansen finally found a way out. In a magazine article, he not only recognized the authenticity of the hairy man, but also confessed that he had shot it himself with an 8 mm Mauser bullet. But this had happened, he said, within the United States, which eliminated the need to explain its illegal introduction into the country.

So, Hansen kept his mouth shut but could not rid himself of the memory of the horrible incident. His conscience also bothered him. Had he killed an escaped gorilla, or had he maybe killed a man, perhaps disguised as some kind of hunting decoy? These obsessive thoughts gave him migraines. He stuffed himself with pills. Unable to concentrate on his work, he tried to avoid flying as much as possible. Finally, he couldn't stand it and decided to settle the issue and to check out the identity of his victim.

Thus, about a month after the incident, on December 3, 1960, following a heavy snowfall likely to leave a clear trail and prevent his getting lost or to wander in circles, he returned to the site, accompanied by his dog.

He started by crisscrossing the area on his swamp-buggy, a contraption riding on DC-3 tires he had built himself to travel in swampy areas. Having found his former camping area, he left his vehicle and, finger on the trigger of his gun, his heart beating wildly, and seized by uncontrollable dread, he followed on foot the path that led him to the clearing.

Searching here and there, he finally tripped on the frozen corpse of the hairy creature. He stared at it in a stupor. So, he had not dreamed this up! He wiped the snow off the frightening head and noticed that "an eye seemed to be completely missing." However, there was so much frozen blood on the face that was difficult to be sure. One thing for sure: although there was no hair on the face, the rest of the body was covered by long dark hair matted with frozen blood. The creature's left arm was twisted under the body. Hansen also noted that its hands were twice as large as his own. Gradually, as he examined the corpse, his fear left him:

> I was now convinced I had not killed a true human being, but something similar to man, perhaps some "freak" of nature. Maybe it was a mutant of some type.

The hairy creature was "perfectly preserved." In contrast, the gutted deer had been completely eaten by predators. Why had they not touched the monster? For sure, thought Hansen, this business was full of mysteries.

Leaving the corpse out in the swamp was clearly out of the question. The first hunter to happen upon it would notify the police, who would eventually find out who had shot it. It was also not possible to bury it on the spot because the ground was frozen solid. Only one solution remained: to take the corpse away and hide it.

The next day, Hansen returned to the scene of the crime, intent on taking the corpse away.

Using a chisel, he cut it away from the frozen ground, taking along some of the ice matrix coating it, as well as chunks of the frozen muck into which it was laying. With great effort, he hauled the ice block onto the platform of his swamp-buggy, tying it up with nylon straps. Back at the truck, which he had used as a carrier for his swamp-buggy, he hoisted the burdensome load onto the back.

After nightfall, he managed to bring his macabre trophy back to his home in the suburbs of Duluth.

Needless to say our adventurous aviator had not reached the end of his troubles. First of all, he had to provide some "explanation" to his wife, Irene, who still knew nothing about all this.[25] Some Sunday! Then, with the aid of his spouse, now duly informed of the circumstances, he had to discretely carry the corpse to the basement, having first sent the three kids to bed. They planned, the next day, to stuff the body into a large freezer, emptied of its normal meat contents. But by Monday, a horrible stench had already invaded the house.

> Despite the stench, we entered the basement and bent the creature's arms and the legs so that it would fit into the freezer. Either the body was still frozen or *rigor mortis* had set in. It was an extremely difficult task and we both breathed easier when the creature was completely in and the top securely fastened.

The idea was that the incriminating corpse would remain hidden until spring, by which time the ground would have thawed and it could be surreptitiously buried. However, Hansen noticed after about a month that the corpse was dehydrating and that some parts were beginning to look like dried meat. He confided to his wife: "If we are to bury it in the spring, it won't make any difference. But if we learn what it is and decide to keep it then it should be properly preserved. I don't know how to keep it from drying out."

Fortunately, as a competent housewife, Mrs. Hansen knew just what to do. She recalled how she had managed to preserve some Canadian trout for two years after freezing them within a block of ice. That was the solution!

That's how the Hansens started pouring 20 gallons of icy water into the cooler every day. After a week, the corpse was completely submerged and entirely embedded in a block of ice.

Spring 1961 soon arrived. Hansen had had time to reflect on the hazards of a clandestine burial. As soon as it would thaw, the corpse would again emit a stench powerful enough to alarm the whole neighborhood. Digging a pit deep and wide enough to accommodate such a large body was bound to attract attention, unless this was done in a completely isolated area. And just imagine that through some unforeseen incident during transport, the corpse—bloody, stinking, and strangely hairy—were to roll out onto the road? Can you imagine having to explain its origin to a probably hostile and scoffing highway patrolman? No, this was much too risky. And, since Mrs. Hansen had by now become used to the idea of keeping this abominable carcass in her freezer, he thought they might as well leave it there.

However, in the summer, the Hansens bought a farm in Rollingstone, Minnesota,

---

25   If this whole story were true, that would have surely been one of Hansen's most difficult tasks!

as a quiet retreat for the soon-to-retire aviator. There was no way around it: they would have to move there with the bloody freezer and its contents. It would have been risky to hand over the move to a commercial firm, which would likely have asked probing questions. So, they rented a large truck and a hoist, and friends of the family helped lift and move the heavy and bulky container. Helpers were told that meat was left in the freezer to ensure that it would not spoil during the trip. Besides, in the chaos of moving, the key had been lost.

The Hansens were enormously relieved when, after seven hours on the road, the sinister freezer was finally deposited out of sight in the farm's shed. The farm was conveniently located in a remote corner of the Midwest, far from neighbors, in an area where one could bury a brontosaurus in the garden without attracting attention.

Given these new circumstances, there was no longer any reason to hurry to dispose of the corpse of the hairy creature. More than four years went by without doing anything about it. Then, in November 1965, Hansen said good-bye to the Air Force, after more than twenty years of active service. Not surprisingly, Hansen soon found the inactivity intolerable. To kill time, Hansen began to read, to nourish his mind—a stark contrast to military life—which was how after a while he became aware of the story of the Abominable Snowman of the Himalayas. And the more he read, the more he began to ask himself whether what he had in his freezer might not be that kind of creature.

In December 1966, Hansen ran into an experienced fairground exhibitor with whom he spoke of the boredom of civilian life. The man told him of the exciting aspects of his trade and encouraged the ex-pilot to exhibit in the commercial show circuit the extremely rare vehicle in his possession: the oldest model of a motorized tractor. It's in the wake of this suggestion that Hansen came up with the idea of also exhibiting the frozen creature, which he thought looked like a prehistoric creature and would be a sensational exhibit at the fairgrounds.

However, before making up his mind, Hansen decided to speak of his plans with his lawyer, a personal friend, to find out what legal risks he might encounter. The lawyer, after an astonished look at the hairy corpse, brought up the likelihood that Hansen might face murder charges, should it turn out that the creature was considered human. He also mentioned a variety of laws forbidding the possession and transport of corpses. Rather worried, Hansen wondered whether he might not, after all, organize just as attractive a show by having a copy made of his "ape-man."

His lawyer found the idea attractive, but remarked: "You have the original body. The authorities will be after it because this thing is the scientific find of the century; however, it might be possible to create a model, as you suggested. Maintain a record of the model's construction but show the real creature instead. If the officials pressure you, it's a small matter to produce photos of the model taken during different phases of fabrication."

"Better than that," Hansen replied. "I'll even exhibit the model for the first year so that it will be accepted by carnies as a 'bogus' show.'"

In January 1967, bringing with him detailed sketches of the original, Hansen went to Hollywood, where he met with Bud Westmore, who was in charge of the make-up department at Universal Studios. Westmore told Hansen that the model he would like to have made would cost him around $20,000. He was too busy to personally take care of the fabrication but would provide technical expertise.

Hansen then contacted Howard Ball, a renowned specialist in the creation of life-size fiberglass animals for museums and similar expositions. Ball was the author of the gigantic reconstructions of the prehistoric mammals placed, for the education of visitors, near the La Brea tar-pits, in Hollywood, where the bones of fossil mammals have been found (giant sloths, saber-tooth tigers, etc.). He was the expert who would be in charge of creating a model of the frozen carcass.

On the advice of another make-up virtuoso, John Chambers, of 20th Century Fox, Hansen met Pete and Betty Corral, who regularly worked for the Los Angeles wax museum; they agreed to implant hair, one by one, into the model.

After he model was finished, Hansen was still worried. He had spent thousands and thousands of dollars on this project, having to borrow much of it, and nothing guaranteed that his exhibit would have the degree of success that could justify such an expense.

> Despite my misgivings, I enlisted the aid of a friend in Pasadena and we added the finishing touches to make it look as close to the specimen in my freezer as possible. The bloody eyes, broken arm, and the blood-soaked hair was carefully duplicated to match the original.

The next phase consisted of embedding the replica in an ice block, like the original, which led to some comical moments. Hansen was expelled from the industrial cold storage warehouse in Los Angeles where he had rented a cold room to store the specimen. On the beautiful sunny day when he showed up with his monstrous model at the back of his station wagon, one of the administrators saw him arriving. Alarmed at the idea that some health inspector might find what looked like some horrible road-kill in a warehouse devoted to the storage of human food, he simply kicked Hansen out with his gruesome cargo.

Hansen ended up making some deal with a small private firm that recently had gone out of business. Properly embedded in a block of ice, the replica was finally lowered, with a small crane, into a refrigerated coffin with a custom-made glass cover. The sarcophagus was then deposited into the trailer, which would serve as an exhibition venue, and driven to Los Baños, California, at the beginning of the West Coast Shows circuit.

May 3, 1967, was a red-letter day: the replica of the hairy corpse was shown to the public for the first time. It was presented by Hansen as a What-is-it?— a complete mystery. All that the public was told was that the frozen creature had perhaps been discovered in that state by Chinese fishermen in the Bering Sea. That was the cover story the ex-pilot had carefully developed and which he planned to stick to for two years.

The circuit lasts from May to November and terminates in Louisiana. During that time Hansen freely revealed "confidentially" to his fairground colleagues that the exhibit shown to the public was only "an artificial creation." Despite the low admission price ($0.35), the exhibit was only moderately successful, since, even according to its exhibitor, "the model contained too many imperfections to fool anyone with an expert knowledge of anatomy."[26] Thus, after returning home to Minnesota, Hansen was more resolved than ever to satisfy to the public's curiosity, as early as the 1968 season, with the real specimen. He slightly thawed both hairy creatures, the real and the false, and substituted one for the other. Referring to the real specimen, he added that: "I worked the creature in a position closely resembling the model by cutting the tendons in the arms and legs. I then started the difficult task of creating ice around the specimen."

From then on, from May to November 1968, Hansen's hairy man attracted much more interest everywhere, not surprisingly, especially at the Oklahoma and Kansas State Fairs. A number of physicians and biologists came, sometimes more than once, to examine the puzzling specimen and to seek information as to its origin. No one, however, dared speak of it in a scientific report or publication.[27]

The rest is well known. It's in December of that same year that Sanderson and I showed up, following a phone call, and paid Hansen a visit to view and photograph, over three consecutive days, the famous specimen whose authenticity no longer left any doubt in our eyes. And it is in February that my preliminary note appeared in the *Bulletin de l'Institut royal des Sciences naturelles de Belgique*, wherein I described the specimen preserved in ice as a yet unknown form of contemporary hominid, to which I gave the name *Homo pongoïdes*.

---

26   In an article which I published before Hansen's "confession" in *Saga*, I had written that as far as I was concerned, the artificial model had never been publicly exhibited because it would not have fooled anyone. Although Hansen's words seem to say the opposite, they do confirm in some way the soundness of my conclusion.

27   Perhaps because of the carnival atmosphere surrounding the exhibit, or also because of Hansen's ambiguous explanations, contradictions, and flagrant variations of his "cover story," which he couldn't keep straight, and perhaps also because of the indiscretion of his fellow carnies, who betrayed his "confidence," as he probably secretly wished they did.

Wrote Hansen:

> My problem started again with the publication of Heuvelmans' article. It seemed as if every newspaper, radio station, every magazine and television station in the world wanted to verify the existence of the creature. Calls poured in every day from London, Tokyo, Berlin, Rome and scores of American cities. The Smithsonian Institution requested permission to inspect the carcass. This request was promptly refused. Dozens of scientists asked for permission to remove a core sample of the creature. Biologists wanted hair and blood samples.
>
> Heuvelmans had stated in his article that it appeared that the creature had been shot. Newspapers began to speculate on the possibility that law enforcement authorities should investigate the manner in which I obtained the creature. "If the body is that of a human being, there is the question of who shot him and whether any crime was committed," an article in the *Detroit News* reported.

Hansen consulted his lawyer, who told him directly: "Frank, if you're not careful, you'll find yourself in jail." He also advised him to replace as soon as possible the real specimen by the model and to take a long holiday. Which Hansen immediately did.

He first arranged to carry out the substitution in a cold room of a food warehouse. He thawed the real hairy man so as to detach from the coffin the ice block in which it was embedded. He then lifted it with large straps placed underneath it, loaded it into a refrigerated truck, and sent it off to a secret location. He then refroze the copy to set it in the place of the original within the showcase coffin.

The final words of Hansen's "confession," chosen with great care, are undoubtedly the most meaningful part of the article, and throw a penetrating light on the whole affair:

> During the past few months, I have been pressed for the conditions or circumstances under which I would consider giving the specimen up for scientific evaluation. Two conditions must be met before I would even consider such an action. One: A statement of complete amnesty for any possible violation of federal laws. Two: A statement of complete amnesty for any possible violation of state and local laws where the specimen was transported or exhibited during the 1968 fair season.
>
> There will surely be skeptics that will brand this story as complete fabrication. Possibly it is. I am not under oath and, should the situation

dictate, I will deny every word of it. But then no one can be completely certain unless my conditions of amnesty are met.

In the mean time I will continue to exhibit a "hairy specimen" that I have publicly acknowledged to be a "fabricated illusion," and leave the final judgment to the viewers. If one should detect a rotting odor coming from a corner of the coffin, it is only your imagination. A new seal has been placed under the glass and the coffin is airtight.

This final paragraph perfectly summarizes Hansen's usual strategy, from which he has never deviated: keep them guessing, maintain the ambiguity. A specimen is henceforth to be shown. The real one? The fake? Maybe yes, maybe no. A specimen that is *presented as false.* Yes, but is it? Finally, it can be anything. A single clue: it is now impossible to detect any odor of putrefaction. But why? Is it because a rubber model cannot rot away? Or is it because, as he says, the coffin is now absolutely well sealed?

What is then one finally to think of Hansen's confession as published in *Saga*? At beginning of the article, the showman declares: "Now, for the first time anywhere, the man responsible for the biggest controversy to hit the scientific world in the past decade presents his own story to the public." However, at the end of that same article, he cleverly back-pedals, saying that everything he said might be nothing but a bunch of lies after all. Since he was not under oath, he was not legally bound to tell the truth. To lie is not a crime. A breach of trust takes place only if one benefits from a deliberate falsehood.

But you might think that Hansen had perhaps been well paid for such a sensational article published in a high-circulation American magazine. Deliberately misleading the public by feeding it a pile of hogwash and being paid for it exposed him to prosecution. Not so dumb! On the very first page, he wrote: "I have not asked for, and will not receive, a single cent from *Saga* magazine." And rest assured that there is no paper trail to prove the contrary. *Everything* had been well planned to ensure that the article was absolutely unassailable from a legal point of view. One can be sure that every word of the story had been carefully screened and had been the object of a critical and vigilant scrutiny by one or more lawyers. That was particularly clear in the last three paragraphs, quoted above.

That being said, such a display of legal precautions will allow me to point out what is pure fabrication in Hansen's story, without being accused of libel. The author himself has admitted his story might indeed be *"a complete fabrication."*

The first time I read Hansen's article, I experienced a feeling of victory mixed with genuine amusement.

I felt victorious because the contents of the article confirmed the soundness of my views on a number of points: except for the Minnesota hunting incident, my reasoning had been perfectly correct about what had happened in the past and what was about to happen in the future. At the same time, I could not help laughing at all the rather transparent wiles, and at the tone of the story, where I could not help but recognize the inimitable style of my friend Ivan Sanderson. The publication of the article in *Saga* was no surprise, as the editor of that magazine, the charming Marty Singer, was a long time acquaintance of Ivan.[28]

Actually, the publication of Hansen's confession looked a lot like a set-up, a scheme born from Ivan's fertile imagination, carried out with the blessing and in collusion with Hansen, and rigorously crafted with the help of one or more lawyers. The aim of the effort was pretty clear. For poor Hansen, it had the effect of sheltering him from legal prosecution, and for clever Ivan, it opened up the possibility of making the specimen available for a complete scientific investigation, and thus to justify ourselves to the world. Ivan had obviously made some deal with Hansen: if he succeeded through the publication of this article in obtaining a formal legal promise of unconditional amnesty for any crimes or offences the ex-pilot had committed, Hansen would then, in exchange, allow Science to examine the original specimen. It was a clever and skillful plan; rewarding if successful, and without untoward consequences if it failed.

If Hansen's story was, by all signs, a lie, it was not necessarily a *pure* lie: it might even reveal much of the truth. Indeed, considering the care with which the article had been written so as to be legally unassailable, it was reasonable to think that *what could be easily verified had to be true*, especially material facts and all the events involving named persons.

Among the verifiable elements, let's note first the "crime weapon." One of the best medical experts in the US confirmed, at my request, that the wounds suffered by pongoïd man could well have been caused by a bullet from an 8 mm Mauser, and that this *kind of weapon was needed* to produce the serious injuries observed: the shattering of the occipital region of the skull and the expulsion of the eyeballs by the power of the shock wave. An 8 mm cartridge, holding a 170 grains (11 gram) bullet, sends it out of the barrel of the Mauser rifle at a speed of 712 meters per second. After 100 meters, the speed falls to 495 m/s, and at 300 m, 420 m/s, which is quite fast! Even at 500 m, the energy of the bullet would be sufficient to cause the damage observed. Unless he was shot with a rifle with a scope, the hairy man was probably shot at a distance of less than 100 meters.

But before picking apart the questionable details of Hansen's pseudo-confession point by point, let's note that what is most striking at first glance is an *absence*: that of the "*real owner*" of the specimen, the secretive Californian tycoon, behind which

---

28   Ivan had introduced me to him on Jan 2, 1959, in New York, when we had lunched together at the Famous Kitchen, an Italian restaurant on West 45th Street, across the street from the Old Whitby, the building where the Sandersons had an apartment.

the ex-pilot had hidden up until then. He is not mentioned a single time; he's not even alluded to. That means one of two things: either the man has never existed, or he wanted to be kept out of a business that now began to stink as much as its hairy protagonist.

The first option is difficult to justify. It is of course possible that the so-called "real owner" might not be whom Hansen claimed he was, a Hollywood character only eager to own objects that others did not possess, but it is certain, as I noted before, that Hansen always behaved as if he had to account to someone else. However, that someone else might well have been what is called a "legal person," meaning a society, an organization, or a syndicate. We shall return to this point when we analyze the financial aspects of this business.

Reading through Hansen's article, one doesn't find any absolute impossibilities or flagrant nonsense—that would have been too clumsy. Everything hangs together and appears coherent and logical. "Never trust an overall impression," said Sherlock Homes, judiciously. "Focus on the details." And as soon as one looks in detail at the facts, a number of weaknesses, huge improbabilities, and even stark contradictions surface.

First, I wonder about the reaction of a military pilot, a career soldier, exposed for more than a decade to the horrors of war, first in Korea and then in Vietnam. He passes out, or nearly so, because he has just shot some kind of a biped gorilla? Even though he is armed, he runs away in a panic from creatures he claims have disappeared and make no move to run after him? He is racked with remorse, enough to lose sleep, to suffer horrible nightmares and persistent migraines, just because he killed, in self-defense, a creature he is not sure is human? He was so disturbed by what was actually only an unusual hunting incident that he no longer dares piloting an airplane? Come on!

I also think about the real miracle—truly a miracle—by which the corpse remained intact after having left out in the open for a month. People have been canonized for less.

It doesn't make any sense that it should have been found "*in a perfect state of conservation.*" In northern Minnesota, the average temperature in October is 7.5°C [45°F], and in November, -4°C [25°F]. The average temperature over these two months *is above zero [32°F]*, and even in November, the temperature frequently rises above freezing, especially during daytime. Normally, putrefaction begins in a human corpse three days after death. After a month at a temperature near zero [32°F] decomposition is *very* advanced.

What is even more surprising is that the carcass of the hairy creature was untouched by local predators, while the deer that was lying near it was completely devoured. Carnivores are abundant in the region: there are wolves, coyotes, foxes, even some bears, raccoons, skunks, weasels, and minks. There are other meat-eaters, like raptors and rats, as well as a numberless legion of invertebrates. Even if the extremely pungent smell attributed to hairy wild men might have deterred some of the most delicate animals, it would certainly not have kept away carrion eaters.

Continuing with the series of suspicious aspects of the story published in *Saga*, I now reflect on the series of steps involved in the transport of the corpse from the clearing in the Whiteface swamp to the basement of Hansen's house in Duluth.

Independent calculations have led me to conclude that the hairy creature must have weighed around 125 kilos [280 lbs]. Such a weight is quite normal for a creature with an anatomy similar to ours, 1.80 m [~6 ft] tall and heavily muscled. (As a comparison, let us recall that an adult male gorilla, only 1.8 m [5 ft 10 in] tall on average, weighs from 140-180 kilos [315-405 lbs].) Furthermore, according to Hansen, the body was encased in a layer of ice because its pelt was thick with frozen blood and it was stuck in the frozen ground, so much so that he had to extract it with a chisel. This layer of ice would probably have added 10 to 20 kilos [22-45 lbs] to the weight of the dead creature.

Now try to imagine a man, even a strong one, trying to hoist *on his own* such a heavy load into a swamp buggy, a vehicle riding on DC-3 wheels and thus rather high off the ground. Have you ever tried to put to bed a sleeping, drunken, unconscious or dead man weighing 100 kilos [225 lbs]? After managing this feat a first time, Hansen does it again by transferring the corpse from the swamp-buggy to the back of his pick-up truck. Finally, with the help of his wife, Irene, who is rather thin and svelte, he manages to carry it down the stairs to the basement. No comment!

Other minor details in Hansen's story also seemed debatable to me. As Sherlock Holmes also said: "It has long been one of my axioms that small details carry the most importance." Rather than engage into theoretical speculations on some of these, I carried out some experiments. Hansen claimed that after a month in the family's freezer, the corpse had begun to dehydrate and that some parts of it had begun to look like dried meat. I did the same: I placed a fresh corpse in a similar freezer. After a month, I had not noticed any change in the appearance of its skin: it looked as fresh as when I had put it in. To be certain, I extended the experiment for six months without observing any such desiccation. I must confess that I did not go as far as assassinating someone to perform this experiment: I used the body of a stillborn mouse, which had the advantage of being hairless and easily observed.

There are other minor improbabilities in the *Saga* story.

For example, why was Hansen expelled from a commercial cold storage plant in Los Angeles when he came in by car to have his model frozen? Even if the plant were devoted to the preservation of human food, why would the presence of a rubber and fiberglass model, which everyone could see was artificial, have disturbed the most fastidious food safety inspector? The horrible appearance of the object should not have mattered: just a step away from the Hollywood movie studios, the operators of cold storage plants in the area would have been used to such model monsters. By the way, it appears that in the United States no commercial enterprise will refuse to deal with you, whatever your wishes might be, if the price is right, the project realizable and within the limits of the law. And even if one oversteps the law, there are possible accommodations made in Heaven.

One also finds in Hansen's account an internal contradiction that slipped by all the participants in the affair.

When Hansen speaks of the finishing touches that he and a friend are working to apply to the replica, he specifies: "*Bloody eyes, a broken arm, sticky hair: everything was carefully reproduced so as to be identical to the original.*" That was absolutely necessary since it had been decided that after a year, the original would replace the model, and if any problem arose, the copy would quickly be substituted again. However, when the time came for the first substitution, Hansen says of the original: "I worked the creature into a position closely resembling the model by cutting the tendons in the arms and legs."

What a mistake! It's obvious that the corpse *already* had—one might say *by definition*—the same attitude as the model built in its image and exact likeness. This is a doubly clumsy statement, because the suggested post-mortem surgery was certainly not necessary. Once thawed, a corpse automatically becomes as soft and malleable as it was before being frozen. A fact I also verified. Every housewife who has prepared frozen meat knows that. The famous *rigor mortis* is a transitory phenomenon which occurs only between three and six hours after death, depending on the environment, and lasts on the average only sixteen to twenty-four hours.

Often in criminal cases, a guilty party who seems to enjoy a perfect alibi betrays himself by adding too many small details aimed at creating realism. Among these apparently negligible and extraneous details there is always one the police discovers to be false. The whole edifice of lies then crumbles. If one has lied on a small accessory detail, might one have not also lied on major points?

Here, for example, is another embarrassing detail. Why did Hansen struggle with his conscience for a month before returning to the scene of the crime? He finally decided to do it on Dec 3, after having waited some time for favorable weather conditions. The hairy man had been shot, again according to Hansen, on the day following the opening season for deer. From what I have been able to find from the Department of Natural Resources of the State of Minnesota, the deer season opened on Nov 12 in 1960. The incident would then have occurred on Nov 13. The month over which Hansen struggled with his conscience lasted only two weeks.

Whatever, the fact is that in spite of the help and skill of Sanderson, and the advice of lawyers, Hansen's brilliant maneuver fell flat. Copies of the article were sent to all imaginable scientific, police, law, and executive authorities without provoking the least reaction. Nobody offered Hansen amnesty from all possible indictments in exchange for a thorough scrutiny of the specimen.

Why not? First of all, in my view, because the confession had been published in a popular magazine, a monthly publication for men, where more credit is given to military or sporting events than to scientific discoveries, wherein one finds photos of juicy pin-up girls rather than half-decomposed ape-men, and where articles focus on the merits of the latest styles of pajamas or sports cars rather than on the most revolutionary theories of human origins. These magazines, at least the best of them, are often very interesting and highly informative about the seamy or adventurous background of various events, even sometimes touching on scientific issues. It is sometimes thanks to the boldness of such publications that important facts, ignored by the scientific establishment, are revealed to the public. However, in this case, as the matter had already been broached in a scientific journal, it would have been necessary, in order to attract the attention of scientific authorities, to continue the discussion in a publication of similar standing.

It is most likely that scientific journals would not have welcomed an article by a fairground stall-keeper, regardless of his status as a veteran, but a science writer like Sanderson, perhaps backed by zoologist or anthropologist friends like John Napier, Carleton S. Coon, or W.C. Osman-Hill, could have commented and critiqued Hansen's version of the facts. (Of course, the story would have had to be boiled down from its dramatic style to a terse and dry scientific style.) I am sure that some American scientific or intellectual magazines, of which there are excellent ones, would have been delighted to cast a fresh light on a matter of extreme interest to the anthropological community. Even if it were only to conclude with a question: "Is this the greatest discovery in the study of anthropology, or an improved version of the Piltdown hoax?" The legal powers-that-be would then have found themselves morally obligated to "do something" to reassure public opinion one way or another. Although, as we shall see, that is far from certain. In fact, *everything* was done at that time to force a reaction from the highest levels of the American administration…all in vain.

Left to right: Carleton S. Coon, W.C. Osman-Hill, John Napier

As soon as the issue of *Saga* containing Hansen's confession had come out, I had myself, on July 6, send a copy to Professor André Capart, the director of the Institut Royal des Sciences Naturelles de Belgique. Capart had had the courage to publish my brief scientific report and had always backed me up in my work on *Homo pongoïdes*, a subject in which he took a personal interest. Adept at seeing through Hansen's story, he discussed its contents with Xavier Misonne, the curator of the Mammals Section of the Institute. Furthermore, as he was a personal friend of King Leopold III, who had also taken a special interest in this mystery from the very beginning, he brought him up to date on the unexpected turn of affairs. He told me about these discussions as follows: "I showed a copy of the article to Misonne and to King Leopold, and both agree with you: this is a made-up story meant to hide the true origin of our Neanderthal."

I was very much comforted by these comments, after all the humiliations and insults suffered since I had the gall to announce the existence on our planet of another kind of human being. And it was much more than mere moral support. For Capart is an eminent scientific personality; Misonne is a reputed mammalogist whose precision and rigor I have always admired; and King Leopold is really not like any other king. His perspective was not simply that of a sovereign rejoicing at the role played by one of his subjects in an important scientific discovery. He spoke as an expert, being himself a keen field naturalist and reputed ethnographer. He is probably the world's most knowledgeable expert on tropical forests, which he has patiently explored in their remotest corners, from Peru to Papua-New Guinea, through the Amazon, Guyana, Central Africa, and Indonesia, and few people are more familiar than he is with the people of all these regions. Quite recently, his enthusiasm for natural history led him to create *Le Fonds Leopold III pour l'Exploration et Conservation de la Nature* (The Leopold III fund for exploration and preservation of nature), bringing together a handful of eminent people among whom I am proud to have some good friends, such as Sir Peter Scott, the founder of the *World Wildlife Fund*,; Professor Jean Dorst, of the Paris Museum [of Natural History]; Walter van den Bergh, the director of the Antwerp Zoo; and of course, Professor André Capart.

Capart, who is a man of action, went beyond just giving me an encouraging pat in the back. In his capacity as president of NATO's oceanography sub-committee, he had met at meetings of that body with Daniel P. Moynihan, who was then scientific advisor to President Nixon.[*] On August 4, 1970, he sent to him, care of the White House, a letter in which he updated the file so as to inform him of the situation and requested in the strongest terms that he do everything possible to allow scientific examination of the specimen by satisfying Hansen's stipulated conditions.

---

[*]   Moynihan was actually the Counselor to the President for Urban Affairs, not the scientific advisor. — The Editor

Truly, it is your responsibility, and I am at a loss for the right words to request that you take advantage of your influence to elucidate this mystery. I assure you that in a similar case, I would do everything possible to discover the truth.

Thanks in advance for anything you might accomplish in the name of Science.

Although America at that time was still gloating with legitimate pride of having put an astronaut on the moon a year before, André Capart had not hesitated in his letter to affirm the comparable importance of *Homo pongoïdes*: "You have there a subject as worthy as reaching the Moon!"

As hard as it is to imagine, this passionate plea remained without an answer. When one of the most eminent representatives of Belgian science sends an official letter to an advisor of the president of the United States, the least that he might legitimately expect is a polite answer. Capart mentioned this to Daniel Moynihan at the first meeting of the CDSM (Comité des Défis de la Société Moderne—Committee on Challenges to Modern Society—a NATO body) when they met again. Moynihan was astounded: he had received neither the letter nor the file that accompanied it. Back in Brussels, Capart sent a second letter to the White House, attaching again copies of my scientific article as well as Hansen's confession. Once more, to no avail. Soon afterwards anyway, Moynihan was replaced and with the loss of the best personal contact with the president of the United States we also lost all hope.

As I had already been through a lot in this affair, I was not too surprised. For quite a while now, all my efforts at finding information or provoking action had run against a brick wall.

In summary, whether because of the context in which Hansen's so-called confession had appeared, or because of its fanciful tone and unlikely events, or because of some occult censorship from above, the fact is that no one in the United States reacted to Hansen's proposal. Nobody moved.

Nobody except Frosty Johnson.

Forrest Johnson, known to everyone as "Frosty" since his youth, was the young director of a painting business in Chicago. He had always been enamored of anthropology, in particular the anthropology of his native Kentucky. He had participated as an amateur in the excavation of tumuli in search of ancient Indian burials. At the university, he had studied chemistry, but after graduation he had discovered a talent for business and had gone into the sale of chemical products. One day, at the barbershop, he accidentally came upon a copy of *Saga* that included Hansen's article.

The story kindled his imagination. If Hansen had really killed that strange hairy creature in Minnesota, one of a group of three similar beings, the other two must still be roaming in the woods, somewhere in the same area. Of course, all that had happened ten years earlier, but such creatures were clearly not the only ones of their kind, and must be part of a local population.

As Frosty was a man of action, he immediately set out to check on a number of points in Hansen's story. He succeeded in reaching Arne Ranta Jr., the son of the owner of the hunting cabin where the other hunters of the 247th Air Wing, based in Duluth,[29] had set up camp in 1960, near the Whiteface Reservoir. Arne did remember the hunting party, but he laughed at the story of the ape-man. In the seventy years that the area had been settled, mostly by people of Finnish origin, no one had ever heard of that kind of creature, or had come upon unusual tracks. And Heaven knows how regularly these forests were crisscrossed by all kinds of folk: hunters and trappers, loggers, geologists, prospectors, surveyors, etc.

These categorical statements did not deter Frosty who, continuing his enquiry, managed to trace and interview one of Hansen's hunting companions, Dave Allison. He remembered Hansen well and with some affection but also as a kind of underachiever—still a captain after fifteen years of service! They all teased him a little about it, but everyone liked him. A strange type, really, always full of off-the-wall ideas. He just couldn't help hatching all kinds of preposterous schemes. His military planning ideas were ludicrous. Nevertheless, it was quite true that Hansen had become lost on that day during their hunting junket. And it was also true that a corpse had subsequently appeared in his freezer. He had even been teased about it. Of course, he was teased about everything.

If Hansen had actually killed some bizarre being, would he, Frosty asked Allison, have been able to bring it back home, stuff it in his freezer, and leave it there for five years? Oh, sure, answered Allison, that was just like him.

This information was enough to convince Frosty of Hansen's sincerity and of the truth of his story. He decided to launch an expedition to the forests of the Whiteface Reservoir area to find these mysterious creatures and to contact them. Before starting, however, he consulted some documentation. He found the article published by Sanderson in *Argosy* and even discovered my own note in the archives of the *Field Museum of Natural History* of Chicago. With this information in hand, he approached the management of the *Chicago Tribune Sunday Magazine* with a proposal for an unusual safari in the swampy woods of Minnesota to verify the existence of some kind of Abominable Snowman now known under the scientific name *Homo pongoïdes*.

The project found the needed support. A professional photographer, Fred Leavitt

---

29   This is by no means a contradiction, as it might seem, with Hansen's description of his unit as presented in *Saga*. I have managed to confirm (with great difficulty, as one would expect, in trying to find out information from the Pentagon on military units in active service) that the 343rd Fighter Group was a fighter squadron belonging to the 247th Air Wing. Its base in Southeast Asia was the Da Nang airport, in South Vietnam.

and a talented journalist, Tom Hall, teamed up with Frosty and his friend, Joe Sheeran, an experienced tracker. Hall went on to report on the results of the expedition on October 24, 1971, in a long article in the *Chicago Tribune Sunday Magazine*, under the attractive title "Tracking the Minnesota Monster."

The account of the expedition itself, of the search carried out by the Finns living in the area, of the systematic sampling of the woods by the four men who were equipped with compasses and walkie-talkies, and of their fruitless efforts to attract their prey with the carcass of a beheaded rooster, ends up being of little interest, except as a negative proof. It turned out that *no one* in the area had ever heard any story about wild and hairy men, not even the vaguest rumor or legend. Nothing at all. After a quarter century of cryptozoological research I can affirm with certainty that in *every* part of the world where an animal new to science has been discovered, its existence was *always* already known to the natives.

It would be futile to suggest that the silence of the local people was due to the obstinate and incredulous nature of the Finnish population, unreceptive to anything unusual or surprising; most of those people reported having seen flying saucers. Some had even seen them land. Some kids told how they had seen three-foot-high beings emerging from them. But no one had ever seen anything that looked like an Abominable Snowman.

Tom Hall's article, wherein all this was presented, did not limit itself to illustrate the unlikelihood of Hansen's hunting prowess, and in particular the inappropriate choice of the northern Minnesota forests as the home of the hairy man. Hall had also consulted with Hansen and Sanderson at length and presented incisive and colorful portraits of both. All this allowed him to present a well-documented and generally error-free account of the events.[30] Had Hall taken the trouble to interview "the third man," the third protagonist of the drama, meaning myself, he would have achieved a rigorous and strictly accurate summary, fully respecting the historical truth. (As for myself, whenever possible, I support my statements with authenticating documentation.) Unfortunately, at that time, I was across the Atlantic, and Hall had perhaps imagined that my version of the events would in any case have been identical to that given by Sanderson if perhaps not that of Hansen. In any case, I could have proven to Hall that he had been misled as to my having given my word to Hansen of never publishing anything about his specimen without his permission, and that in doing so I had broken a solemn promise. I have on hand no less than

---

30   There was a single obvious error in detail, difficult to understand: Why did Hall give 1965 for the first appearance of the frozen man on fairgrounds, when it took place exactly on May 3, 1967? One might believe it might be a typo but for some other chronological reference in the text that seems to imply the same erroneous date. Other minor errors are of much less importance and do not affect the basic soundness of the story.

four documents—two reports, a letter, and even a published article—which prove the contrary, all written and signed by one of his informers.[31]

What was of great interest for me personally in Tom Hall's article was that he confirmed the accuracy of some of my conclusions and the relevance of some of my suspicions.

First of all, Frosty Johnson's enquiry showed that it was unlikely that Hansen had killed a hairy man in Minnesota.

Furthermore, Hall revealed that the *Saga* article in which Hansen had confessed his murder had indeed been written by Sanderson. Hall relates that shortly after the publication of Ivan's article in *Argosy*, Hansen had approached Sanderson. "This time, he brought an article of his own pen. It was, according to Sanderson, a rather stupid piece of writing. Sanderson edited and re-arranged it in his own style, rewrote it for Hansen, and undertook to have it published in some other men's magazine."

Finally, among the confidences Hansen shared with Hall, there reappears, larger than ever, the shadow of the famous "owner" of the specimen, strangely absent from the confession published in *Saga*.

"I am not the real owner," suddenly confessed Hansen. "There is someone else, very rich and very bright who likes owning that sort of thing. He paid for the truck. He paid for the model. All that cost more than $50,000; I would not have had the means to do it. I call this person the owner of the specimen, although I did not really sell it to him. We only have a verbal agreement and he has the upper hand. I do what he wishes. He doesn't want anyone to know who he is."

Hansen adds that it's the rich eccentric, his "sugar-daddy" sponsor, who advised him to keep the corpse out of the indiscrete probings of the scientists and told him to exhibit it in fairgrounds. It was he also who picked up the original specimen in a secret cold storage when it had to be moved. His men took it away in a refrigerated truck.

The tycoon, adds Tom Hall, had now hidden the Thing in a secret location where no one will ever find it. If Hansen ran into trouble again, all he would have to do is to make a phone call. The monster would be loaded on an eighty-foot yacht and dumped to the bottom of the Pacific Ocean.

Hansen claimed to have suffered a great deal because of his loyalty towards that powerful man, but not even the worst Chinese torture would make him betray him. "Hansen was ready to put up with the anger of the whole world to be a shield for his

---

31   Without referring to the more confidential documents, I shall limit myself to quoting from a published document, an extract from an editorial from the April 1969 issue of *Pursuit*, the bulletin of the Society for the Investigation of the Unexplained, led by Sanderson: "Doctor Heuvelmans and the director [Sanderson] went to examine the specimen and immediately recognized it for what it was. However, Hansen demanded that the latter [Sanderson] should not publish his discoveries unless he had an explicit permission for the owner itself, who was said to be 'an extremely influential but eccentric West Coast personality.' Doctor Heuvelmans did not make any such promise."

sponsor. He had to, of course. His sponsor would do the same for him. Only once, when he wrote that famous confession, had he done something without consulting the tycoon. That had been pure stupidity."

Overall, Hall's article provided an excellent overview of the whole story of the frozen man; it also filled in some small missing pieces of the puzzle, which was far from completely resolved. From now on it was possible for anyone, based on the articles written by Hansen, Sanderson, and myself, supplemented by articles published by some well informed journalists, to reconstruct the sequence and relationship of events and most importantly to provide an unassailable proof of the authenticity of the highly controversial specimen.

Chapter 7

CORPSES COME AND GO BUT ARE NEVER THE SAME
Behind the scenes and the proof of the specimen's authenticity

"As the story of the Minnesota Iceman unfolds, it turns out to be a problem for a detective agency, rather than for a biologist."
— John Napier, professor of primate biology
at the University of London, in his 1972 book, *Bigfoot*

Everything I have reported so far has been verified or can be checked without difficulty; a special bibliographical index provides information on published references and I have in hand unpublished documents in case anyone is interested. Of course, statements made by individuals only reflect the views of their authors and bear no guarantee of veracity, but the fact that they have been expressed can always be verified, whether they were made in front of witnesses, or put on paper and signed, or printed with the permission of the author, or registered on tape in the case of telephone conversations.

In the light of the known facts and their careful comparison, one might begin to venture some conjectures on various obscure points of the affair and attempt to answer some of the following questions:

What role did Terry Cullen play in the sequence of events?

How much revenue did exhibiting the frozen man bring to Hansen and his potential associates?

Who is the legal owner of the specimen?

It would then be easier to answer the more fundamental question:

Is the original specimen authentic?

Or, more precisely, since for someone like me who has examined it most carefully and carried out a prolonged anatomical analysis there is no doubt as to its authenticity,

Is it possible to prove that the original specimen is authentic?

Once that is proven, there will only remain a few questions pertaining to the legal aspects of the issue, which will clear up by themselves. A god-sent sensational development will come about at the end to justify my conclusions.

But let's proceed systematically and first try to understand the character of the first person who, through his phone call to Ivan Sanderson, first launched this fantastic adventure.

As soon as one thinks about it, it becomes obvious that Terry Cullen had to be in cahoots with Hansen. He had to have been very close to him to have the opportunity to see the hidden backside of the specimen. To lift a 400 kilogram [900 lbs] block of ice (an operation which implies some defrosting so as to free it from the walls of the freezer in which it was lying) is a long and difficult task that is not undertaken to satisfy the curiosity of just any of the thousands of visitors. If Cullen actually saw the underside of the specimen, it is most likely because he participated in its placement into the glass-topped sarcophagus, a ceremony in which only Hansen's closest friends would have collaborated. If he weren't there, Hansen would have told him what he had missed seeing had he been there on that occasion. Why? So that he would tell Sanderson about it, of course.

It is indeed in order to attract Sanderson's attention, to make his mouth water, so to speak, that Cullen spoke of a sagittal crest, which does not appear to be present on the specimen. The fact is that such a cranial structure is usually attributed to the Himalayan Snowman as well as to the Californian Bigfoot by all those who, like Sanderson, have attempted to study and describe them. Mention of such a characteristic trait was bound to attract Sanderson's interest. Cullen's competence in herpetology also offered some assurance of the validity of the information. So, Cullen's phone call and the confirming letter, followed by the promise of finding out the address of the holder of the specimen, were clearly aimed *at attracting Sanderson to Hansen.* The latter had also not been fooled by Sanderson's attempt at justifying his visit through a pretended commercial interest: Hansen already had at home copies of Sanderson's articles on Bigfoot; he had also heard of, and perhaps even read, his book on the Snowman.

Just why Hansen wanted to meet Sanderson remains a matter of speculation. It was certainly not for publicity, at least not immediate publicity; Ivan had much too much difficulty in obtaining permission to speak about the exhibit. The most likely explanation is that the ex-pilot was seeking the advice of the most famous American expert on the Snowman about the real or potential nature of his own specimen. To me, it seems that Hansen was particularly interested in finding out from his visitor *whether it might not be possible to claim that the creature, originally from Asia, might actually have been shot in the United States.*

In that case, Hansen would have some hope of forever escaping from the sword of Damocles suspended for years over his head: the risk of having to explain some day how the hairy corpse had fraudulently been brought into the United States.

Anyway, later events were to confirm this hypothesis. Like it or not, the only benefit that accrued to Hansen from his meeting with Sanderson was that the latter, thanks to his knowledgeable assistance in the planning and writing of Hansen's "confession" in *Saga*, had rendered more acceptable the idea that the pongoid man had originated in the United States.

Let's now turn to the financial aspects of Hansen's carny show, which are far from negligible. Indeed, we have witnessed the unfortunate carny endure, because of his frozen man, many problems, torments, and fear; we have seen him having to continuously adopt new strategies; we have seem him struggling to escape from a difficult, even dangerous predicament while disdainfully rejecting generous financial offers. It's really legitimate to ask: was it worth it for him?

Without assuming anything about the authenticity of the specimen, and without taking into account its initial purchase price, one can nevertheless reach some estimate of the financial status of the "Frozen Man" operation.

In his article, Hansen stated that, in the words of Hollywood make-up specialist Bud Westmore, fabrication the replica would cost around $20,000. Other experts estimated its cost to reach at least $15,000.

The process of inserting hair into the rubber or polystyrene model would by itself have cost $3,500. Calculations demonstrated that this is not an exaggeration. I counted at least 17 hairs per centimeter square on the specimen's chest, an area where the hair density is normally lower than on the back, the top of the head, the neck, or the thighs, but also higher than on the face, the palm of the hands, or the soles of the feet, as well the armpits and knees. Let's assume a uniform hair density distributed over the whole body surface. For a man 1.80 m [5 ft 9 in] tall, that area is about 2 m². To cover such a model would require approximately 340,000 hairs. At the rate of 10 hairs per minute, each at its own individual angle, one could stick about 600 hairs per hour in a rubber model. For a single person to implant 340,000 hairs would require 566 hours, i.e. twelve 48-hour weeks: three months of boring and meticulous work.[32] A fixed price of $3,500 for the whole operation would imply an hourly wage of $6.00, which is rather low in the United States for such delicate specialized work.

The price of more familiar pieces of equipment required for his show is much easier to calculate.

The total investment cost can then be estimated at least at $50,000 as follows:

Trailer and tow-vehicle $30,000
Freezer... $5,000
Model $15,000

---

32   Of course, the work could have been done in a month-and-a-half by two people, and in three weeks by three workers.

As to the annual revenue, Hansen provided an estimate that corresponds to the maximum theoretical value, claiming that at 35 cents per person admission to his exhibit would bring in $50,000 per year. This amount corresponds to about 150,000 visitors, a maximum number difficult to reach. In six months of the show season, this would translate to 25,000 visitors per month and thus about 1,000 per day (with one day off per week). What this means is an uninterrupted sequence over an eight-hour day of groups of 10 persons spending less than 5 minutes looking at the coffin holding the specimen.

In three years, the business should have brought in at the most $150,000 gross. In practice it probably brought in much less, maybe no more than $100,000. One should also remember that Hansen thought that the first year of his show had not been particularly successful, perhaps because of the imperfection of the model, which would have led him to exhibit the real specimen, as early as 1968.

Nevertheless, let's go with Hansen's estimate of gross income. To find out the net income, we have to subtract operating costs (permits, space rental in fairs, housing, gas and vehicle maintenance, insurance, etc.) perhaps more than $10,000 per year and thus $30,000 over three years. One should also subtract the non-redeemable items (the artificial model as well as the sarcophagus, which do not really have a resale value), as well as the depreciation of the vehicles. It quite certain that after three years of operation Hansen would have been left with used vehicles and less than $100,000.

We also know, from Hansen himself, that he did not have the means to finance such an operation, which is consistent with his rather modest lifestyle. How could he, at the end of his military career as a captain and father of three children, have saved $50,000? He clearly had to borrow that amount, as he admitted, and he thus had an associate (at least one!).

In this kind of partnership, involving at least two participants, one providing the funds, the other the work, it is customary to share the benefits 50-50. In this case, each partner would have ended up, after the operation, with $50,000 (before taxes) plus half the value of the trailer and tow vehicle. This means that the funder would only have recovered his initial contribution plus half the resale value of the vehicles. As to Hansen, he would have worked for three years to earn the same amount, that is $17,000 a year, not much in the United States. He would also have the right to half the resale value of the vehicles.

All the above, we must remember, assuming the maximum gross revenue. We have not even included the initial cost of the original specimen. As a comparison, these days, one can buy a live gorilla for about $6,000. Contrary to what one might imagine, the price of an orangutan is not much more, even though it is an endangered species and its trade is strictly prohibited.

Overall, the "Frozen Man" business was not, and couldn't be even *a priori,* much of a moneymaker. For Hansen, and Hansen only, it was a normal business that did not justify taking serious risks. It is very important to know this if one is to discuss the identity of the person who might be the "real owner" of the specimen.

Even if Hansen's exhibition had probably been financed by a third party, it's not impossible that he might have been the owner of the original specimen, which he first said to have purchased in Hong Kong and then claimed more fantastically to have shot in Minnesota. However, in both versions, different and contradictory, of the origin of the hairy corpse, he always insists that there is another person who has rights on the specimen. On the one hand, he had admitted to Ivan and myself that "someone" had lent him money to purchase it from the Chinese merchant; he had even repeatedly added that this someone was the "real owner." On the other hand, he was to reveal to the newspaperman, Tom Hall, that although his powerful "patron" was not the legal owner of the specimen, he had rights to it and had the power to dispose of it as he pleased, even to the point of jettisoning it to the bottom of the sea in case of serious troubles, which all comes to the same thing. Anyway, as I pointed out in the first chapter, Hansen had inadvertently and spontaneously let out in his automatic reactions that he indeed had to account to someone else.

So, is the owner or patron simply, as Hansen always claimed, a wealthy and original Californian wishing to own things that nobody else does, a movie mogul who either bought on a whim or had Hansen purchase the specimen for its weight in gold, or a Hollywood nabob who could afford to refuse colossal offers and could set up a private laboratory for the study of his oriental curiosity by sympathetic experts?[33]

My answer to that question is a definite NO! When one is so possessive of something, one doesn't leave it to the crowds to gawk at for three years. When one believes that the specimen is of incredible scientific value, one does not exhibit it in county fairs. And when one is so rich that even an offer of a million dollars is rejected out of hand, one is also not interested in a business that can only bring problems and at the end the nuts and bolts of half a trailer and its tow-vehicle.

---

33   As soon as Hansen began to describe the "real owner" of the specimen, anyone familiar with American life couldn't avoid thinking about Howard Hughes, the millionaire Texan born in 1905, aviation pioneer and famous Hollywood producer, just as well known for his whims and extravagant tastes as for the secrecy of his private life. As he had not been seen in public or photographed in the press for a dozen years, there circulated all kinds of outrageous speculations about him: for example, that he was dead but that his passing had been hidden so as to prevent the collapse of his business empire. Actually, Hughes is very well, thank you. The latest news was that he was in England and was thinking of buying property on the island of Jersey. However, just for fun, every time someone asked me to retell for the thousandth time the story of the frozen man and claimed in triumph to have guessed the identity of the specimen's owner (Howard Hughes), I had gotten into the habit of reply: "You are so right! Actually, it is Howard Hughes himself who lies in the coffin. His body became covered with hair from continually hiding in dark places. After his death, he was frozen so as to be able to pull him out in case there was any problem with his inheritance…" I am surprised that no one took this explanation really seriously. [Howard Hughes died April 5, 1976. —Editor]

It is obvious that this anonymous person—or as I suggested earlier perhaps a corporate person or association—cannot be involved in this carny fair just as a dilettante or an eccentric collector, or as a generous benefactor, or by mere financial interest. Any relationship with Hansen can't simply be on the basis of a straightforward business agreement; there must be between them some understanding, possibly based on some long-time friendship or debt of gratitude. Or, as one says, might there be "a skeleton in their closet," not only in reality but also figuratively? Not only the carcass of some hairy hominid, but joint participation in some illegal activity?

Why would this person, or corporation, have helped Hansen to the tune of $50,000 to finance his exhibit without expecting some reasonable profit? Why did that person remain in the shadows and never show up, even in the more critical moments? Why is he no longer in the picture in the "confession" published in *Saga*, the article in which Hansen requested total immunity as the price of a scientific examination of the specimen? Before that article (and even after its publication, in his revelations to Tom Hall) Hansen had always ascribed (and continues to do so) all responsibility to that mysterious person.

Those are perhaps the most mysterious aspects of this affair, to be resolved only in the light of its legal status. Those aspects are obviously linked with the question of the authenticity of the specimen. It cannot be the same laws that are broken when the specimen is a flesh and bone creature than when it is only a rubber dummy!

As for myself, I am of course perfectly convinced of the specimen's authenticity, and so is Ivan. But is it possible to communicate to someone else a conviction based on personal experience? Is it possible to prove formally that we have not been misled?

In his book entitled *Bigfoot*, Dr. John Napier, who never took the trouble to examine the frozen man, although that would have been easy for him, devoted a whole chapter to defend his view that it was a hoax. One has to say that Napier finds himself in an embarrassing predicament: he bears the main responsibility for discrediting so offhandedly the strongest evidence for the greatest anthropological discovery of all times. The history of science will never forgive him.

What troubles Napier most is that "two experienced zoologists like Heuvelmans and Sanderson could have been misled." His only explanation is that we were troubled by a nightmarish atmosphere worthy of Charles Addams, under which we examined a "brilliantly crafted model."

I am quite ready to admit that I have in my life been *very* afraid, and I can even specify when: as a child, I found myself one night in a small boat that threatened to capsize because of the waves produced by a steamer; again, as a young man, when I found myself stuck in a narrow tunnel during a caving expedition; and finally, as a soldier during World War II when my anti-aircraft gun was strafed by Stukas. I was also on occasion very impressed by being charged by a black rhino, or a furious elephant; swimming in waters infested by sharks and barracudas; photographing at close range mortally venomous snakes. Those are all memorable experiences. I also have to say that I am less afraid of the most dangerous animals than I am of people, especially hostile crowds. But I have to say that this did not keep me from walking alone at night in the streets of Harlem, or to go dancing in Africa or Central America in places where I was the only European, or to confront hoodlums in Mediterranean slums. I also know what it is to be awakened early in the morning during the German occupation by officers of the Sicherheit Dienst (Security Service) who did not like what I wrote, and also paradoxically to be thrown into jail at the Liberation by small-time resistance thugs, armed to the teeth and thirsting for violence, who did not like the very same writings. I have known war, captivity, persecution, many marriages, and supposedly incurable diseases. I have sometimes been shaken but never, even in the worst circumstances, to the point of losing my wits. Even if were merely denying it, it is rather childish and grotesque to claim that a frozen corpse, in a fairground trailer, would have made such an impression on a zoologist who has spent a good part of his life working in museums full of skeletons and jars of horrible specimens, or dissecting in the laboratory all kinds of animals, even sometimes eating his lunch among the gently breathing bodies of anesthetized animals, guts hanging out of their body cavity. That also holds of course for Ivan, whose life was perhaps even more adventurous than mine and who had tamed many more mammals, and especially monkeys, than I had.

In spite of all that, I am ready to admit, if anyone wishes, that for three days in a row I was shaking with terror, except when I took my photos, since they are all in focus. Had that been the case, it would have made no difference. I have already said that if the fabrication of a fake was impossible *in practice*, I have to admit that it was *theoretically* possible. To me, it is only possible to establish the authenticity of the specimen for someone who did not examine it (or who would have done so under emotionally disturbing circumstances) through circumstantial proofs or by a demonstration *ab absurdo*. And what could possibly be more rigorous than a proof by geometry, as commonly used in rigorous science.

Before appealing to geometry, let's note that if the specimen were a fake, the whole affair would just be a gigantic hoax, entirely motivated by a desire for profit. It does not make sense to imagine an attempt at hoaxing or ridiculing anthropologists, as was the case for the Piltdown Man hoax. Had that been the objective of the counterfeiters, they would for sure have staged the discovery in a more serious setting than a fairground, and would not have enlivened the story of the specimen's origin with the kinds of tales told by Hansen.[34]

If we suppose that the original specimen was a fabrication made for profit, a plethora of questions immediately come to mind, many of which cannot find a sensible, logical answer. Anyway, I have not been able to find anyone who could provide such answers. I would be willing to give in if my opponents were to offer some answers rather than hiding behind under a veil of blind incredulity, without any other arguments than "it's impossible," or "it's a mystery," or even "the hoax hypothesis is in any case the most reasonable."

IF THE ORIGINAL SPECIMEN IS A FAKE:

1. *Why did Hansen try to belittle it?*

To convince others that a fairground exhibit is interesting, one should also be, or appear to be, interested. But until the day he withdrew his specimen from public view, Hansen never claimed, whether in private or in public, that his specimen was authentic. That might have been because he didn't want to be accused of having earned money through false advertising. But Hansen did not even suggest that the specimen could have been authentic; for him, it was some kind of oriental fabrication. Alternately, he pushed the idea that it was a very ordinary creature, claiming that analysis had shown that its hair belonged to a well-known Asian type (Ainu?) and its blood was perfectly normal (an unusually hairy *Homo sapiens*?). Starting on April 20, 1969, Hansen deliberately advertised his exhibit as a fabrication. It is only in September of that year that he began to suggest that the specimen was authentic, four months before withdrawing it from public view. Why?

2. *Why did Hansen advertise his exhibit with such ridiculous labels?*

To ensure that a fairground exhibit will be taken seriously and be perceived as an enigmatic item likely to stimulate the curiosity of the crowd, it is expedient to present it with a certain degree of decorum, preferably in some pseudo-scientific jargon. That is a time-honored trick of charlatans. Having conversed with him and carefully perused his library, I know that Hansen is not an ignoramus. Lacking neither wits nor a certain veneer of culture, he surely knew the difference between "illusion" and "allusion," and was well aware that medieval men did not live in glacial times. Why

---

34   For example, they could have released the ice block in the middle of winter on an Alaskan beach, making sure that they could have retrieved it after having warned the press, who could have taken pictures, and invited a few scientists to observe the specimen before cutting the block from the ice floe. The only problem would have been to find a good reason to take away the specimen to make sure it wouldn't be submitted to a careful analysis.

then those absurd and misspelled advertisements, likely to fool only illiterate yahoos. Why?

3. *Why would Hansen shy away from widespread publicity?*

The commercial success of a fairground show is linked to its fame. However, at the beginning, Hansen strictly forbade Sanderson to write about the exhibit in *Argosy,* a magazine with a readership of nearly two million. That's really difficult to understand since the proposed article aimed to authenticate the specimen and emphasize its scientific interest, a move likely to attract the crowds. Why would anyone forgo an opportunity to enhance its commercial success? Why?

4. *Why was the specimen decomposing?*

An artificial model, made of rubber or of some plastic material, has been shown to have been constructed in Los Angeles. Why then did the specimen that Sanderson and I examine stink of putrefying flesh? That could happen if it had been constructed, as Sanderson suggested, from the corpse of a chimpanzee. But in that case, why have a copy made of that fake? Why?

5. *Why was Hansen so embarrassed when we detected the smell of his decomposing specimen?*

The smell of decomposition could only help establish the authenticity of the specimen. Why then did Hansen, far from being happy, blanch and appear surprised, and later annoyed, when Sanderson pointed out the smell and had him check it for himself? Why?

6. *Why did Hansen reject attractive purchase offers?*

Since Hansen had never claimed, before his "confession," that his specimen was authentic, he could sell it, even if fake, as a curiosity of unknown nature without being accused of fraud. I calculated that his business possibly returned a gross maximum of $150,000 over three years, and after costs, a net profit of $50,000 a year. He received or heard of potential offers of up to ten times that amount. Why did he reject them all? Why?

One might imagine that such refusals might have been motivated by a wish to double his profit by first benefiting from the exhibit and then from the sale of the specimen. But once the circuit was completed, far from trying to sell the specimen, Hansen made it un-saleable by declaring it authentic (at least in case, as we suppose here, that it was a fake). In that case, he exposed himself to being accused of fraud by selling it after such a declaration. Why did he not try to sell it before this "confession"? Why?

7. *What problems could be avoided by creating a replica of the specimen?*

Hansen admitted (as was verified) that a replica was made earlier, at great expense, "in case one day problems arose." What kind of problems could there be if the specimen was a fake? Certainly not a charge of fraud, since it had never been presented as authentic. Hansen had only suggested the existence of an original, an authentic creature, but he had never claimed that this was what he was exhibiting. Why then speak of potential troubles? Why?

8. *Why did Hansen repeatedly refuse to have his specimen x-rayed?*

One can understand why before March 20, 1969, Hansen would have refused to have his specimen x-rayed if it was a fake; his exhibit would have lost much of its mysterious aura and hence its attractiveness. But from the moment that he publicly announced that it was a fake, ridiculing all the specialists who had claimed it was authentic, as well as the Smithsonian and the FBI, he would have enhanced the interest in his exhibit and the comical impact of his revelation by showing authenticated x-rays of a rubber or plastic specimen, most likely mounted on a wire mesh carcass.

Furthermore, in July 1969, when he was stopped at the border crossing from Canada to the United States, and was under suspicion of having possession of a human corpse and transporting it across the boundary, why did Hansen chose to spend 24 hours in this awkward situation, and to spend a lot of money on long-distance phone calls, to disturb lawyers, a much more expensive proposition, and even to solicit the intervention of a US senator, rather than proving his innocence by allowing his specimen to be x-rayed? The owner, whoever he might be, would have been none the wiser. Why such stubbornness? Why?

9. *Why did Hansen continue to change his cover story and finally end up denying it entirely?*

If the whole business of the frozen man is one giant hoax, why didn't Hansen, right from the start, invent a plausible and coherent story to explain the origin of the specimen, a story that he could then have stuck with without change? But either in person or through Terry Cullen, he offered at the beginning at least three different versions of the discovery of the specimen (a Russian, a Japanese, and a Chinese). He also first claimed to be the owner of the specimen, before attributing its ownership to someone else. He also sometimes related that it was he who had found the specimen in Hong Kong, sometimes that it was the owner who had found it, and some other times that it was his fly-boy friends who had told him about it. Finally, after insisting for three years on an Asiatic origin, he had changed his tune and claimed a North American origin. Why this final about-turn, which, if true, turned all the others into lies?[35] An inveterate liar certainly doesn't inspire confidence and these continuing contradictions do not jive with the idea of a clever hoax. Why so many blunders? Why?

10. *Why did Hansen suddenly confess to having murdered his specimen?*

If there never was an authentic specimen, why did Hansen confess to having killed a creature, which he recognizes as human-like, having hidden the corpse for ten years and having transported it repeatedly across state lines and even across international boundaries? All these are criminal activities; the first one is a capital crime. Of course, under American law, statements not made under oath are not incriminating. However, such statements nevertheless attract the attention of the law. What serious reasons led Hansen, who was until then wary of attracting police attention, to suddenly expose himself to the possibility of being closely watched and likely to be arrested for the least misstep?

It can't be for a fee resulting from the publication of the article in *Saga*; Hansen swears that he didn't get a penny for it. It can't also be for the purpose of raising the selling price of the false specimen since his confession has the opposite effect of devaluating it. To avoid being charged with fraud, one can only sell a fake at the price of a worthless curiosity.

The only possible explanation for Hansen's confession of the "murder," assuming that the specimen is indeed a fake, might be that it enhanced its value as a fairground attraction. That is possible, at least in theory, but he couldn't really benefit from it. It is illegal to seek profit via fraudulent means. Thus, if the *Saga* article were to include fallacious assertions, it would become a crime to receive money for its publication (which Hansen knows well since he refused to be paid for it). For the same reason, Hansen cannot ask for a viewing fee to look at a fake specimen if he has claimed it to be authentic. Before his "confession," he could exhibit the hairy corpse as a "total mystery"; subsequently, he can do it only if he publicly declares it to be a fake. (That was indeed his intention, as he stated at the end of his article.) So why then this spectacular "confession," which can only bring trouble? Why?

---

35 I must quote here from John Napier's book about Frank D. Hansen: "I don't believe he ever lied: he only answered each question evasively… He never claimed that his exhibit was anything else than a mystery, which it was—and continues to be." No comment!

11. *Why would Hansen demand a prior promise of amnesty before even considering the possibility of yielding the specimen?*

Before his confession, Hansen had never claimed that his specimen was authentic, and if his specimen was a fake, his activities were perfectly legal. There was nothing to fear. So what crimes were in need of an amnesty? Why this insistence? Why?

12. *Why was the specimen decorated with wounds?*

I already raised this question in Chapter 2, noting that while bleeding wounds add a theatrical aura to the exhibit that is likely to please the public, one of these wounds, namely the absence of eyes, takes away a spectacular and unique effect: the gaze of man presented as a remote ancestor, a look from the darkest past. If the specimen was manufactured, why forgo such a powerful feature? Why?

13. *Why doesn't the specimen look like the traditional reconstruction of a prehistoric or fairy tale man?*

If the hairy man has been fabricated, what model was used to design it? In fact, it doesn't look like any of the classical or traditional reconstructions of Neanderthal Man, be they scientific or artistic, nor, one should note, of the various Australopithecines or Pithecanthropes. It doesn't even look much like the best sketches of the Himalayan Yeti or the American Bigfoot. Actually, its extreme hairiness, crooked toes and upturned nose resemble only a little-known reconstruction of Neanderthal Man put forward in 1922 by Dr. Maurice Faure at the Montpellier congress of the French Association for the Advancement of Science. It is only in a reconstruction by R.N. Wegner, considered a caricature by anthropologists, that such an upturned nose has been given to a Neanderthal.

As I already stated in Chapter 2, had the specimen more closely resembled the generally imagined figure of a prehistoric man or ape-man, it would have met with a lot more success on the fairgrounds and might not have been completely ignored by science for a year-and-a-half. So if it was indeed a fabricated specimen, why wasn't it given a more "commercial" appearance? Why?

Many other questions remain desperately unanswered, even unanswerable, if Hansen's exhibit is only a hoax. We will be happy with the above thirteen questions, which cover most aspects of the affair. The consequences of these questions are incomprehensible in the case of a hoax. It is thus completely absurd to suppose that the original specimen was a fake. *Thus, it is authentic: Q.E.D.!*

As a means of verifying this conclusion, we will see later that it is easy to answer all the above questions if the original specimen is truly a real corpse. However, let's first clear up the legal aspects. Hansen's concerns, maneuvers, precautions, hesitations, and changes of mind clearly seem to indicate a fear of judicial pursuit. But for what crimes?

I first raised this question in Chapter 4, but from a purely theoretical point of view. Now, in the light of the ups and downs of the affair, and having determined the authenticity of the specimen, we will stick to the two distinct versions offered by Hansen for the origin of the specimen: Asiatic or North American. Here is then a list of the possible legal infractions arising in both cases:

(1) The specimen is a human being, at least in the broader sense of the term, so having killed it could be considered homicide, a capital offense.

(2) To hold a human corpse without informing local authorities and without special authorization from State and Federal authorities is a crime.

(3) To transport a human corpse without such permissions across State or international boundaries is a federal crime.

(4) Possession of scientific or artistic items of great value must be declared and such items must be registered.

(5) It is illegal to bring anything into the United States without declaring it, even if there are no customs duties to pay; smuggling is a federal crime. Transporting contraband items from one state to another is also a federal crime.

Finally, since Hansen was an Air Force officer at the time when the specimen was found or brought into the States—and was perhaps smuggled in by him from Asia—one should also note that using military equipment, such as an airplane, to commit a crime is a matter for a court-martial and subject to severe punishment.

As one can see, there is no lack of infractions of varying severity for which Hansen had the best reasons to want to hide at all cost the authenticity of the specimen.

According to his first version of the facts—an Asiatic origin of the hairy man—Hansen would have been guilty of the last four (2, 3, 4, and 5) infractions. According to the other version—the North America origin—he would have been guilty of the first four (1, 2, 3, and 4). In both cases, he would have been guilty of the same three (2, 3, 4), which are the least severe. What this implies is that by confirming the second version, through his confession in *Saga*, Hansen chose infraction 1, rather than 4, i.e., preferred admitting to murder than to smuggling.

That might at first sight appear to be a rather stupid decision, but upon reflection it turns out to be quite clever. It's clear that no jury in the United States would think of condemning Hansen for having shot a terrifying creature, vaguely resembling a gorilla, aggressively confronting him. In contrast, the illegal introduction into the States of a rotting corpse and the likely use of a military aircraft for that purpose would be extremely costly. One can then understand that Hansen would have preferred to replace a highly plausible version of the facts—the purchase of an exotic curiosity in the Orient—with the improbable and inconsistent but much less inculpating story of a murder. It was also very clever of him to insist after his confession for a "*total amnesty from all possible infraction the law*" without specifying exactly which law might have been broken.

What is unquestionable is that laws were indeed broken, that Hansen was afraid of being indicted for having broken them, and that he hoped strongly for amnesty.

Now, we can easily answer the thirteen questions that remain unanswered if the fairground exhibit was a hoax.

IF THE SPECIMEN WAS AUTHENTIC:

(1) Hansen belittled his specimen to make people believe it was a fake because proving its authenticity would expose him to serious legal problems.

(2) For the same reason, he advertised it in a rather grotesque fashion up until the day he withdrew it from public view.

(3) Hansen first avoided too much publicity for fear that scientists would take his specimen too seriously, threatening a detailed examination that could have revealed its authenticity.

(4) The original specimen is rotting away because it is a real corpse, not a rubber or plastic dummy.

(5) Hansen was afraid that the smell of putrefaction would betray the authenticity of his specimen. He also feared that his exhibit might deteriorate too quickly.

(6) Hansen and his associate(s) or patron(s) could not sell their specimen because the new owner would have discovered its authenticity. One supposes that the legal implications of such a discovery would have been dire enough to reject a fortune.

(7) If Hansen and his cronies had an artificial replica made of the original specimen, it was to avoid threats of legal action following discovery of the authenticity of the specimen. Just showing the proofs of the manufacture of that copy (invoices, photos at various stages of construction, statements by the artisans, etc.) would be enough

to deter further indiscretion. Hansen quickly realized the utility of this maneuver, designed, as he admitted, to avoid troubles and to make believe that there perhaps existed only an artificial model, and to discourage me or the leaders of the institute with whom I am affiliated from publishing a scientific note on the specimen; such a publication was bound to lead to a request for a deeper examination of the specimen if it did not suffice by itself as proof of the authenticity of the specimen.

(8) Hansen consistently refused to allow the specimen to be x-rayed, and even more so to have a biopsy taken. That, of course, would have established without doubt its authenticity (as well as its identity and extraordinary value) with all the ensuing legal consequences. Hansen did not hesitate to contradict himself repeatedly regarding the origin of the specimen, first of all since for that part of the story preceding its purchase in Hong Kong it was easy to determine that the story had entirely been made up (which I did by establishing that the specimen had been frozen artificially). There was an advantage to letting that part of the story pass for a series of tall tales told by the merchant. Indeed, it probably seemed impossible for Hansen, at that point at least, to invent a story to account for the existence of the specimen that would be coherent, precise, and detailed (but not verifiable) without by the same token revealing its authenticity. It was then preferable to remain imprecise until the right opportunity arose. That was probably what motivated Hansen to consult Sanderson (via Terry Cullen). He would then seize the opportunity through the article published in *Saga*.

(9) Rather than being a rigorous and masterfully orchestrated scheme—which would "deserve the Barnum prize, if there were one," as John Napier claimed—the whole business appears to have been a laborious improvisation, initiated by an accident and reworked as circumstances dictated. Hansen and his accomplices believed, wrongly, that the fabrication of a replica would provide definitive insurance against any trouble. If they had been better informed about the status of wild or hairy men on a global scale and had, right from the beginning, stuck with the story published in *Saga*, or some similar but more sensible story locating the death of the creature in Bigfoot country in the Pacific Northwest, they would probably not have been challenged. They would have been able to freely advertise the authenticity of their specimen, which they could have proven by x-rays and analyses; they would have crisscrossed the United States in a triumphal and lucrative tourney and would undoubtedly have easily received permission to exhibit the specimen abroad, and finally, tired of wandering from fair to fair, they could have finally sold the specimen to a scientific institution for at least a million dollars. I doubt very much that anyone would have listened to me if I had objected that the exhibited creature, with its wide and short feet, could not have left those enormous, long and narrow footprints found in Oregon, California, and Washington State.

(10) If Hansen admitted to being responsible for the death of his specimen within the United States, it is simply because this new and improved version of events allowed him to avoid serious legal problems (smuggling and improper use of a military airplane) and perhaps could also open the door to possibly selling the specimen for its real and considerable value.

(11) If Hansen demanded a guarantee of amnesty, it was, of course, to have a solid guarantee of freedom from legal indictments.

(12). The specimen has wounds because it had really suffered from violence, either before being killed by a bullet in its face, or as it was riddled with bullets, one of which broke its arm before another hit it in the right eye.

(13). If the specimen's external appearance does not conform with the traditional reconstructions of prehistoric hominids, in spite of a careful analysis that reveals it to be a Neanderthal, it is simply because the current idea of the appearance of these creatures is mostly speculative (Chapter 3) and often illogical, contrary to what we know of the biology and anatomy of Neanderthals.

As we can see, the answers to all thirteen questions are perfectly coherent and logical when one supposes that the specimen is authentic.

There is only one weak point, as follows:

Are the penalties for the illegal introduction into the United States of the corpse of an unknown victim, even using a military airplane, serious enough to explain why the specimen was hidden for six years and why Hansen panicked upon publication of my scientific note, and why he, or the real owner, wouldn't even consider an offer of a million dollars for the specimen? Most crooks would risk years in jail for much less than that! Would it not be worthwhile to spend a few years in the clink knowing that one could then be wealthy for the rest of one's life?

There is indeed a mismatch between the severity of the infractions and the importance of the precautions and maneuvers deployed to avoid exposure on the one hand, and on the other, the magnitude of the financial benefits eschewed.

Both Ivan and I noticed that disequilibrium early without consulting each other. It was obvious to us that the frozen man had been slipped into the US through a well worn, probably much more lucrative conduit, which could also lead to many years in jail.

In any case, in his counter-memorandum of June 19, 1969, to Dr Napier, Ivan Sanderson wrote about the illegal importation of the specimen that:

In view of Hansen's time in the military, this could have been achieved only with military equipment. Where it really stinks however is that this is the drug smuggling highway, and the West Coast gangs have been battling the Mafia for decades to control the local supply. The owner(s) made a mistake, as happens to the cleverest crooks, in getting involved in the carny business with, or for, one of their men without imagining for a moment that this might provide a clue to their really lucrative activities. We are now talking about big-time crime!

Besides that, whatever they brought back from China, I believe they didn't know what it was and didn't give a damn anyway. So, when Hansen asked if he could have it to make a little money, they just said: "Go with it."

In my own counter-memorandum to Napier, dated the following day, I had nearly simultaneously suggested that the agreement linking Hansen with the mysterious Hollywood mogul might be connected with some former involvement in the smuggling of drugs or young Asian prostitutes. I wrote that: "It is quite likely that Hansen's persistent reluctance to reveal the real origin of the specimen arose from his fear of also revealing the existence of the smuggling network."

How did Napier react to these suggestions by Ivan and myself? Here's what he says in his book *Bigfoot*:

Another rumor was that the Mafia, which might have some unspecified interest in this affair, was exerting its considerable influence to prevent the scientific investigations which I was undertaking on behalf of the Smithsonian. The cloak and dagger school of thought, which sees an iron fist in every silk glove, was loose, and I have to admit that for a couple of days I gingerly walked around Washington with a firm grip on my British umbrella.

We already know how, at that time, Napier had convinced the Smithsonian to abandon its interest in this affair and, in his press release, had irreparably discredited it. Here's how he justified his attitude:

It is possible that the story of the Ice Man might go beyond the patter of a carny, and that it might be more complicated. Perhaps it involves sinister harmonics such as smuggling, secret societies or some kind of racket, but if that is the case, I can't recognize them and I don't care. Cloak and dagger is not my trade: it is biology, and my arguments are based on biological probabilities.

Let's maintain a modest silence on the nature of "biological probabilities," which resulted from the anatomical study of a clumsy sketch of *Homo pongoides* by Sanderson and on the measurements he took using a method of debatable accuracy. Let's simply ask ourselves the question: Where does a biologists' trade end when he is pursuing his research? Is it below his dignity to become a world traveler, like Darwin, or a digger, like Leakey, or an alpinist, a spelunker, or a diver, as many have? Is it even possible to become a great biologist if one doesn't have the soul of a detective?

Anyway, stepping out of the confines of Napier's "good little biologists," Ivan and I wrapped ourselves in our grey cloaks and, dagger between our teeth, patiently continued our investigations, each on our own, occasionally and in spite of our differences, exchanging messages of encouragement across the Atlantic.

As for myself, one specific point puzzled me: *how actually had it been possible to bring the specimen into the USA?* It seemed to me that answering that question would throw light on everything.

At the very beginning, when Hansen was still saying that at the time of its discovery the hairy man was found within a 6,000 pound block of ice, I immediately imagined the problems involved in the transport of such an object, let alone its introduction into the country past the US Customs officers. A later study of the ice itself, of its fine structure and of the shape of its opaque areas, convinced me that the specimen had actually been frozen in a domestic freezer and that the original ice block probably didn't weigh more than 1,500 pounds (see Chapter 10).

The problem with transporting this enormous block of ice lies not only in its weight. It is also necessary to maintain the block at a sufficiently low temperature to prevent it from thawing. The simplest method would be to transport it within the freezer used to freeze it, powered by batteries.

Even then, it is difficult to imagine how it would have been possible, without being noticed, to bring such a package into an airplane, and bring it out upon arrival, to carry it in and out of at least two military airports without having to pass through a number of checkpoints.

Thus, I was led to believe that at some point the pongoid man had been transported alive after being sedated by appropriate injections. That would have reduced to a minimum the problem of bulkiness and also had the advantage of facilitating the covert introduction of the specimen into the United States. I spent time in 1945 in American military camps, where I was assigned to Special Services, and I am familiar with their prevailing atmosphere, rather faithfully portrayed by Robert Altman in his movies *M.A.S.H.* and *Catch 22*. I can easily imagine how one could have brought the inert body of the hairy man out of a military airport, either by dressing him up in a G.I uniform and passing him for a buddy drunk on Bourbon, or by playing chummy with the guard: "Come on, Jack, let me through with my ape. It's only some kind of orangutan I am bringing back from over there. Don't give me trouble, I only want to bring a surprise birthday gift to my old lady…"[36]

Believe me, it's easier to bring a gorilla out of a military compound than bringing in a pretty blonde. And still….

The only problem with the hypothesis of the specimen brought alive is that it does not account for the mismatch between the lightness of that infraction and the scale of the maneuvers that took place over the next ten years to hide the fact.

So I came back to the idea that it was indeed a corpse that had been brought in from the Far East. It is not impossible that it might have been found in a block of ice from the very beginning. I recalled that Hansen had let slip during a conversation with Sanderson that the first time he had laid eyes on the specimen in the cold room of a Hong Kong merchant the strange creature *was inside an enormous bag of heavy plastic.* Surrounded by enough dry ice within that envelope, the ice could have gone a long way without thawing. All one had to do was to enclose it within a box larger than a coffin. Actually, why not in a coffin?

My friend Jacques Paoli, a reporter for *Europe I* and *R.T.L.*, who is always well up on everything happening in the world, once subtly pointed out to me that since there already exists an illegal flow of corpses from the USA to China, there is no reason that there might not also exist a traffic in the other direction. It seems that many wealthy Chinese emigrants to the United States wish, for religious or sentimental reasons, to be buried in their homeland and that they spend big money to ensure that their body is covertly taken there after their death. That led me to think that many dead bodies are regularly brought back to the United States from Southeast Asia: those of soldiers killed in Vietnam.

Would it not have been possible to slip in the carcass of a hairy man among the bodies of those poor boys being returned to their families?

Just to be sure, I enquired about the manner in which these things were done. Here are the results, as they were communicated to me (one will understand that I cannot reveal the names of my informers in military matters):

> The bodies of dead soldiers are repatriated by air from various airports in Vietnam. The most important in the south is that of Ton Sunut. In the north, that of Da Nang. In the central high plateaus, there are two: Chu-Lai and Chu-Chi. The most important airport for this kind of operation is that of Da Nang.
>
> Dedicated aircraft carry the bodies to the United States, but it frequently happens that a cargo plane takes off with only a few

---

36   Everyday, animals whose transport is strictly forbidden, such as the orangutan, are slipped across borders under a plethora of aliases. One can't insist that every border guard should also be a zoologist.

corpses on board or as few coffins as room allows. These aircraft are not refrigerated—the cargo bay is very cold. Before leaving Vietnam, the bodies receive military honors and are embalmed. C-130s are the planes normally used for this transport. Those bodies which are not too damaged or mutilated are placed in coffins which may be opened. Those soldiers badly torn to pieces or reduced to a mush are first put into a heavy plastic bag and then placed into a sealed coffin labeled DO NOT OPEN. Corpses already decomposed are treated with a hardening compound (a chemical product causing stiffening and tanning) and put in a plastic bag and sealed in a coffin labeled DO NOT OPEN.

That Da Nang airport was precisely the base of operations of the 343rd Fighter Group, Capt. Hansen's unit, might of course be just a coincidence. Furthermore, that Hansen's first glimpse of the pongoid man was in a heavy plastic bag might also be a coincidence. The interesting point was that it appeared quite possible in a Vietnamese military airport to hide an odd corpse—or anything else for that matter—in a coffin labeled DO NOT OPEN with absolute confidence that its content would not be subjected to further scrutiny.

I also could not help thinking that such possibilities must have given ideas to persons interested in smuggling goods whose traffic ensures astronomical profits. We might recall at this point what Hansen had hinted about the kind of commerce that the Hong Kong import-export specialist who had sold him the specimen was carrying on "a little bit of everything…from marijuana to something more or less serious…" And how about the parallel drawn by Sanderson between the usual itinerary of Hansen and his flight crew members across the Pacific and "the drug highway."

I had now realized that this "highway" was also the road over which were transferred "untouchable" coffins. And I also recalled another of Holmes' maxims: "When you follow two separate chains of thought, Watson, you will find some point of intersection which should approximate to the truth."

So I felt that I had finally discovered the missing piece in my puzzle or, more precisely, to have discovered where it would fit. I had put my finger on the means by which pongoid man was most likely to have been brought from Asia, which usually served for the importation of rather more lucrative goods than a rotting Neanderthal.

I am certainly not of the school of cloak and dagger, as suggested by John Napier. On the contrary I hate melodrama, as I have clearly shown in all my books; I always try to rely on the simplest explanations for fantastic hypotheses. It is indeed from that perspective that, as soon as possible, I gave a dry scientific name—*Homo pongoides*—to the hairy man, to avoid dramatic or ridiculous monikers such as "Abominable Snowman," "Whiteface Ape-Man," "Minnesota Monster," or "Bozo." But this time,

I really had the feeling that I had stuck my nose in something extremely dangerous.

At the beginning, I found it rather strange that my various suggestions, discretely and tactfully presented, to urge the American authorities to look into an affair, which stunk figuratively as well as in fact, were consistently ignored. Already in July 1969, Ivan had told me that in the United States, he had been politely asked by some bureaucrats to "drop it." As for me, I was told in veiled words or by eloquent silences that I was not minding my own business and that I could rest easily; the appropriate authorities had the affair well in hand.

It is quite true that looking into smuggling and the arrest of drug peddlers is none of my business, as Dr Napier pointed out. But to try every possibility to gain possession of a specimen of inestimable value to zoology and anthropology is indeed my business. Of course, it is not the business of those gentlemen of the FBI, the US Customs Service, the Drug Enforcement Agency, or the Military Police, who couldn't give a damn about Neanderthals. I have to add, however, that I have had among my correspondents some cops—if I may so call them in a friendly fashion— who have sometimes offered brilliant suggestions of a zoological nature to help in my research. It has never occurred to me to tell them to mind their own business; on the contrary, I warmly thanked them and congratulated them. After all, the collaboration of specialists of different disciplines often brings remarkable results.

The fact is that all my efforts remained, at least in appearance, completely ignored. However, one day, on December 16, 1972, my attention was drawn to a news item on France-Inter radio that set my pulse racing. It was revealed that for years, *the corpses of soldiers killed in Vietnam had been the means by which large quantities of heroin from Thailand was introduced into the United States.*

This time, I had found the missing piece of the puzzle!

Chapter 8

CLOAK AND DAGGER
Why the deck was loaded to start with

"Power, like a devastating pestilence, pollutes whate'er it touches."
— Percy B. Shelley, "Queen Mab," Act III

"Something is rotten in the Kingdom of Denmark."
— William Shakespeare, *Hamlet*, Act I

What had happened? The American media told us, but didn't say much about it. At least, they did not keep it in the limelight for very long. Perhaps they had already said too much for some.

On December 11, 1972, a military cargo plane flying in from Bangkok landed at the Andrews Air Force Base, near Washington D.C. There were 62 passengers on board as well as the coffins of two soldiers killed in battle. A man wearing the uniform of a Sergeant First Class walked down the gangway, wearing his medals and decorations attesting to at least ten years of active service, the perfect image of a Vietnam War hero, the fearless knight in shining armor.

He was not greeted by enthusiastic applause, but by the Military Police and by agents of the US Customs Service who asked him for his identity papers. He was Thomas Edward Southerland, age 31, from Castle Hayne, North Carolina. His military documents had been issued in Baltimore for the 18th Airborne Brigade, based at Fort Bragg, near Fayetteville, in the same state, and had been signed by an officer named Ben Jones. His assignment had been issued by a military hospital in Bangkok to allow a certain Captain Paul E. Moe an R&R trip in Southeast Asia.

A careful interrogation and some verification soon established that the so-called Mr. Southerland was a disguised civilian, a truck driver by profession and a sometimes concierge at the Elk's Lodge Pavilion in Greensboro, North Carolina. His military identification was either counterfeit or obtained fraudulently.

During that time, the aircraft on which Southerland had arrived was submitted to a thorough search by US Customs officers who, interestingly enough, paid particular attention to the coffins of the dead soldiers and even to their corpses. They did not find anything out of order. At least, no drugs. Which apparently left them puzzled.

Nevertheless, suspected of illegally wearing a military uniform and in possession of false ID documents, Southerland was arrested and incarcerated.

It later turned out that in this case, the government agents had missed the mark.

Informers had warned them that upon departure from Thailand a 44-pound package of heroin had been placed inside one of the corpses being brought home. The aircraft was supposed to land in Dover, Delaware, but the FBI had diverted it to Andrews AFB, Maryland. However, in the meantime, probably during a twenty-four hour layover in Hickam AFB, in Honolulu, the heroin package had been removed. The federal agents presumed—at least that's what they said—that a preliminary search of the aircraft during refueling in Okinawa had led to removal of the drugs, suddenly found "too hot."

During the enquiry on the Southerland affair, an FBI agent, Joseph Stehr, appeared as a witness in front of Judge Clarence E. Goetz, but his main testimony related to the fact that the accused was "not a member of the Armed Forces." Southerland's assignment paper was signed by an officer who either did not exist or was not authorized to sign such papers. The supposed assignment was not in his name. Some newspapers noted that only civilians were implicated in this affair.

What emerged from the federal agent's testimony is that the FBI was aware of the fact that a contraband organization in Southeast Asia was regularly shipping heroin to the United States, as packages worth $20,000 were hidden inside the corpses of soldiers killed in Vietnam. The packages were picked up either at Fort Lewis (Washington) on the west coast, or at Dover AFB (Delaware) on the east coast. Southerland was part of the operation as a carrier; his role in this disastrous (for him) trans-Pacific flight was to accompany two corpses, one of which was stuffed with heroin. So when he stepped off the plane, Southerland, along with another passenger whose name was not revealed, was detained and interrogated, leading to his imprisonment. For reasons quite different than expected, however.

Judging that there was sufficient evidence to keep him behind bars for a number of reasons, such as false identification and impersonating army personnel, Judge Goetz set Southerland's bail at $50,000, a considerable sum for such infractions.

On January 2, 1973, Southerland appeared before a Grand Jury in Baltimore. He was indicted under nine charges to have impersonated a sergeant of the US Army, to have used false mission documents and official identity papers, and to have made false declarations to customs officers. If found guilty, he faced a possible sentence of twenty-seven years in jail.

In the indictment, there is no mention of drugs or smuggling. However, when the defense requested a lowering of the amount of bail required, Michael E. Marr, the adjunct federal prosecutor, objected to this act of clemency, repeatedly reminding the court that the affair is a conspiracy already linked for eight years to the introduction of "large quantity of heroin into the United States from Southeast Asia." The prosecutor specified that the drugs were "most often" placed in sealed plastic containers hidden inside the bodies of dead soldiers following their autopsy by military physicians. Sometimes, he added, the drugs were simply put inside the sealed coffins of some war victims.[37]

Needless to say, the attending reporters wanted to know more. But the government attorneys objected, saying that the affair was the object of a "top secret" enquiry. Here's what the *New York Times News Service* reported on Jan 3, 1973, in a posting from Baltimore:

> Spokesmen for the United States Attorney's office here, the Federal Bureau of Investigation, the Customs Bureau, the Bureau of Narcotics and Dangerous Drugs, and Army and Air Force intelligence have uniformly declined more than perfunctory public comment on the Southerland case.
>
> But privately one official said today that "it could be deduced" that agents, informed of the smuggling plan, had "gambled" on making an arrest in the United States "with full evidence" and had lost.

One official close to the investigation added that there might well have been some bungling, and he promised that "a continuing investigation" would "produce results."

On the same day, the *Washington Post* went a step further and stated that it had heard from "well placed sources" that the enquiry was nearly complete and that it was likely to lead to the arrest of a dozen people.

Six months later, there was still no word on the results of that enquiry.

Nevertheless, one should note that this was an extremely important drug issue. According to official accounts of the Bureau of Narcotics and Dangerous Drugs (BNDD) for 1971, the Golden Triangle, or Three Boundary region, where Laos, Thailand, and Burma meet, shipped out more than 750 tons of opium and provided by itself *more than half the amount of illegal drugs sold in the whole world.* According to Robert H. Steele, a member of the US Congress, these numbers are underestimates; his own sources suggested that 1,000 to 2,000 tons are exported each year from the Golden Triangle, i.e. 70 – 80% of the illegally consumed opium on the planet. It goes without saying that the American government agencies were loath to recognize the importance in today's drug trade of the opium originating in Southeast Asia, since it is an unfortunate consequence of the already unpopular war in Vietnam. In the United States itself, at the end of the Second World War, there were only about 20,000 heroin and morphine addicts. At the beginning of the 1960s, when the Vietnam War reached its peak, there were already about 50,000. In 1969, the number reached 250,000; in 1972, it had already attained half-a-million and it was expected to reach 800,000 at the beginning of 1973.

Of the 750 tons grudgingly recognized by the BNDD as the illegal production of

---

37   Of course, those were the special coffins used for bodies that had been ripped apart or had decomposed, and were labeled DO NOT OPEN.

the Golden Triangle, 100 tons went, according to that agency, to supply the American soldiers in Vietnam, Thailand, Okinawa, and the Philippines, as well as users in the United States proper. It's obvious that the portion of this enormous amount consumed by the GIs and their friends on the spot can't be more than a small fraction of the whole. The illegal market within the United States consumes 10 tons of heroin a year, which is precisely the amount that can be refined from 100 tons of opium.

Like it or not, there can no longer be any doubt that the major part of the heroin sold today in the United States no longer originates, as it used to, from the Near East and India, transiting through Turkey and then through France and Germany, but now comes directly across the Pacific from the famous Southeast Asian Golden Triangle.[38] The heroin reaches its destination through a series of steps which have for years managed to escape the eye of federal agents, whose efforts have met with very little success, judging from the terrifying increase in the number of addicts in the United States. As one might expect, these efforts usually lead to the arrest of active or retired military personnel, a rather unsatisfactory situation for a government at war. For example, in 1971, a man named William Jackson, who had served with honor under the Stars and Stripes, was arrested. After his military service, he had settled in Thailand, where drawing on the help of his former servicemen still on active duty, he began to organize a regular transport of heroin towards the homeland. The drugs were quite simply placed on board military aircraft or on Navy ships.

To use for criminal purposes coffins and even the corpses of dead soldiers being sent home was the height of refinement in this practice. The elimination of such a macabre shipping method should have been a decisive battle in the America's war on drugs.

The FBI and US Customs agents involved in these operations are neither amateurs nor jokers. They are trained veteran operatives, expert in their specialty. They do not gamble, and strike only when certain. They never just casually launch an operation that's likely to end in a dead end since it can't be tried again; once a smuggling ruse has been found out, it's never tried again.

How is it possible then that their first try, which was to be a masterpiece, should have flopped so miserably?

Why, for example, did they not place one of their own within the suspect aircraft? He would have then been able to prevent a superficial search in Okinawa, caused perhaps by some overzealous agent, or to transform it on the spot into a very thorough search. He would also have been able to counteract the disastrous consequences of that clumsy control and, in Honolulu, would have kept close watch on the coffins rather than leave them without surveillance for twenty-four hours.

---

38   The BNDD has already admitted that 30% of the heroin consumed in the United States originates in the Golden Triangle.

Actually, it is difficult to believe that it could have been the unexpected search in Okinawa that alerted the smugglers. Once the search completed without finding anything, they should have felt more confident: such bad luck can't happen twice during the same trip. It's much more likely that someone warned the smugglers of the planned interception. But who? Who beyond the ranks of government agents knows about such operations?

And how can one account for the fact that after having been warned, the smugglers prudently withdrew their drugs from the airplane, but did not tell their man on board to make himself scarce? Southerland, with his medals and false documents, was much too obvious. Why was he made a scapegoat? It was quite careless to sacrifice to the police a man who might know too much and might be willing to spill the beans.

In the end, one feels like having witnessed a game where nobody followed the rules. In that game, the behavior of the thieves is just as incomprehensible as that of the police.

Southerland's case also raises a lot of questions. How could a mere civilian have managed to obtain counterfeit Army documents—or rather real documents given to the wrong person—and succeed, during wartime, to enter and exit an area of military operations?

That was really a major achievement. The list of supporting documents needed by an American serviceman on overseas duty, especially in a combat zone, is quite impressive: an ID card validated by an intelligence agency, a health booklet listing his medical history, various mission statements, and most important, the Personnel Record, a document which must be on hand at all times. These documents are issued by different departments, and it's extremely difficult to gather them all fraudulently, since that would require accomplices in each of the separate departments or else some extremely influential sponsors.

This story of heroin-stuffed corpses reeks of the same stifling atmosphere as that of the hairy corpse exhibited on fairgrounds.

Whatever one thinks of this affair, whichever end one tries to grasp, and whichever corner one digs into, one always runs into the same implied interference from above, a veil of silence, or quasi-heavenly intervention.

Why the hell is all that "top secret"? Why do the spokespersons of all official agencies suddenly become mute when asked for clarifications? Why is it that whenever, thanks to a new incident or fresh consequences, some explanation is glimpsed, is it immediately hidden again, leaving us with an announcement of some enquiry that never sees the light of day? It's as if one was watching a bad striptease, boring by its slowness, disappointing for its failed promises. Every time Truth reveals the tip of a breast, with the promise that she will soon appear naked, the waiting drags on interminably, beyond the patience of an angel. Finally, when one is ready to give up, a trumpet announces the next number, which turns out to be just as disappointing.

In my opinion, these gentlemen backstage have no intention of showing us their star, the Truth, in all its simplicity. Not that they are jealous. It is because they are ashamed, for it is not nice to look at.

This story started with a weak smell of putrefaction emanating from the cracks of a poorly refrigerated coffin. It is going to end up on the shores of an ocean of stench spreading over many continents.

The explanation for Uncle Sam's prudishness has to be found, I fear, in the subtle complications of high-level politics.

First of all, one should be aware that in 1972, according to the BNDD itself, 80% of the opium trade in the Golden Triangle was controlled by the KMT, the Kuomintang, the Nationalist Chinese Party in power in Taiwan.

Following the victory of the Chinese revolution in 1949, the leftovers of Chiang Kai–shek's army, beaten and expelled from Maoist China, nevertheless remained on the Asian continent. After having regrouped and reorganized, these troops appeared to the CIA as a useful instrument of anti-communist resistance. They were used by the CIA from 1951 to close off the Sino-Burmese border in order to prevent the penetration of Maoist elements into Southeast Asia and perhaps even to attempt to cross the border into Yunnan and recover part of the Chinese territory. After having sought to settle into Burma, an effort thwarted by a UN intervention, the armed units of the KMT ended up working for the Thai government and were an important contributor to the anti-guerilla operations against the communists in the mountainous north of Thailand. By 1961, abandoned by their brethren in Taiwan and by the Americans who no longer needed their services, the KMT bands, numbering around 6,000, began to organize the opium traffic on a large scale in Thailand, as they had previously in Burma. They became the refiners and exporters of the drug, and eventually acquired quasi-exclusive control of this industry in Southeast Asia.

In the meantime, however, the Vietnam War had taken on bewildering proportions, and the Americans, once again, needed the help of all their natural allies: Thailand and its KMT clients, Laos, and of course South Vietnam, and had to demurely close their eyes to the lucrative business their friends were engaged in. One should specify that in the midst of the confusion accompanying the war, it was now military leaders of the friendly forces who personally took care of the drug traffic. Besides the old soldiers of the KMT like General Twan Chi-wen, commander of the 5th Army and General Li Wen-hwan, leader of the 3rd Army, there were, to name only the most prominent, General Ouane Rattikoune, chief general of the Laotian army; President Phao, head of the Thai police; and in South Vietnam, President Thieu himself, whose electoral campaign was entirely financed by the heroin trade.

And that's not all! It seems that in the whole of Southeast Asia, the State Department and the U.S. secret services did not merely ignore, for political and military reasons, the trade in drugs that caused havoc with the soldiers in the field, and also fed the domestic U.S. market. They had to choose between two wars: one against communism, the other against drugs. They chose first to fight the Red Menace and to deal later with the evil poppy. Thus, sometimes they went as far as facilitating the drug traffic to help their allies and allow them to finance their own war efforts. It was even said that the CIA itself had participated actively in the drug trade for a number of years, at least until 1968. This fact, among others, was revealed by an American academic, Alfred W. McCoy, in his book *The Politics of Heroin in Southeast Asia*, published in 1972. It goes without saying that the CIA tried its best to prevent publication but couldn't show that it contained false or slandering allegations. As to the State Department, incapable of countering McCoy's report, it cynically took the tack of stating that his accusations were *out of date*.

I am not inventing anything, and you shouldn't imagine that I have access to secret sources that would allow me to unveil a carefully camouflaged scandal. All this is well known. All this has been published already. I am only quoting facts that the American press, in its great freedom, has reported, facts that have led to reports to Congress by some American politicians, or that have been revealed by former CIA agents, or admitted by high-level military officials.[39]

This takes us very far, I am sorry to say, from the fascinating problem of the survival of Neanderthal man, very far even from that pitiful frozen man exhibited in fairgrounds all around North America. However, without being aware of even the more grandiose antecedents of the latter, it is impossible to understand why it has been subjected to such a blackout, which has resulted from the disappearance of what was undoubtedly the most-important-ever specimen of interest to anthropology. That is a scandal, the Watergate of Science, and I am the one to proclaim it, as I am likely the only one able to do so.

The resulting scientific tragedy is probably the fortuitous outcome of unfortunate circumstances. The hairy man had, by all evidence, been introduced into the United States by taking advantage of a means normally used for the importation of large amounts of drugs. On board military airplanes, rather ironically, heroin accompanied the dead hero until, either by blunder or neglect, the Wild Man took the place of its eternal antithesis, the Knight. To acknowledge the authenticity of the former and its real origin implies recognition of the macabre network. That would have annoyed a lot of people. That's it; it's as simple as all that.

The Neanderthal's funeral cortege wandered off onto the most dangerous road in the world. At its beginning, in Southeast Asia, the warlords of a number of countries set up, for the love of money, a historically unprecedented poisoning enterprise which

---

39   I cannot possibly go into all the details of this sordid business. I refer the French reader who would like to know more to the well documented work by Catherine Lamour and Michel R. Lamberti, *Les Grandes Maneuvres de l'Opium* (Editions du Seuil, 1972).

for many years the State Department and the intelligence agencies supported out of diplomatic necessity.[40] The use of military aircraft from Vietnam to the United States implies the collusion of many within the Army and Air Force, perhaps at very high levels. On the other side of the world, in the US, the drug traffic is controlled by the Mafia, with its summary ways and nearly unlimited financial resources. Thus, at each step of the path leading from the Golden Triangle to the small-time American street peddler, one is likely to run into extremely powerful interests, be they strategic, political, or financial. That type of encounter is often fatal.

Within the scope of such a formidable scheme, what can possibly be the importance of the jarring corpse of a Neanderthal whose identification might, even for just a moment, disturb the functioning of the machine? At best, a grain of sand in the gears, easily disposed of with the flick of a finger. In a business with a bottom line of billions of dollars in the U.S. alone, how much weight can a museum piece worth a million dollars carry, especially if it's an embarrassment?

Even if one dislikes cloak and dagger stories, one is forced to believe when one witnesses someone falling with a dagger in his back and then sees the police throw upon it a cloak to hide it from onlookers. What I have seen murdered in this fashion and then hushed away is a first-rate scientific discovery. Whoever might be involved, I can neither accept nor forgive.

Since I love justice as much as the truth, I must make clear that I am not accusing anyone in particular, and certainly not Hansen, who seems to me more victim than culprit in this business. It is ridiculous to imagine him as a fearsome drug trafficker. Hansen is a simple man who lives a modest life, certainly not like someone grown wealthy from this kind of trade. He also appears to be a man who may have stuck his finger in dangerous gears and, once caught, knows not what to do to extricate himself.

Here's how I see things. My version of the facts is, of course, based partly on conjecture, but it fits perfectly with what is known and certainly shows that the facts can be explained without too much sophistry.

---

40   One may well wonder how far that influence was effective beyond the war theater, perhaps still for diplomatic reasons. Anyway, on June 29, 1973, John E. Ingersoll, director of the BNDD resigned with much fanfare, publicly accusing the White House of having interfered in his operations. Ingersoll said that he was resigning because he was fed up with these continuing interventions, but also because he had been asked to find some other employment by two of the closest personal collaborators of President Nixon, H.R. (Bob) Haldeman and John D. Ehrlichmann, both of whom were subsequently fired for their role in the Watergate affair. Mr. Ingersoll revealed that previously he had been approached almost daily by another former presidential assistant, Egil Krogh. However, he said that the White House did not attempt to influence the course of operations or the hiring of personnel, but he added that this was only because he had clearly signaled that he would not tolerate any pressure of this kind. When asked by the media to explain what kind of interventions took place, Ingersoll only said that "It's difficult to explain. It's the kind of thing that one has to have experienced on a day to day basis."

One day, either by luck or on the basis of a tip, Hansen found himself in Southeast Asia in front of the corpse of a bizarre hairy man that somehow struck his fancy. Without having any idea of what it was, imaginative and likely to elaborate extravagant projects as he was, he immediately understood the benefit to be had from exhibiting such a creature. However, how was he to bring it back into the United States?

Certainly not through official channels. The Army would surely not allow it, perhaps merely for sanitary reasons. Hansen told his buddies about it, probably one of them in charge of repatriating dead soldiers bodies smelled a profitable deal or perhaps just wanted to help the sympathetic captain. One can imagine him taking Hansen aside and whispering out of the corner of his mouth: "Listen, Frankie boy. We'll get your hairy gorilla across; it'll be child's play for us. But there is one absolute condition: not a word! Whatever happens, you know nothing, absolutely nothing. And if you were ever to talk, you would be in deep trouble, the biggest trouble that can happen to anyone. You know what I'm talking about, eh?"

With or without financial implications, the deal is done. Perhaps Hansen never actually knew how his specimen made it to the United States. He received a favor and he is now bound by an absolute rule of silence. It's now a matter of life or death for him to keep quiet.

There still remains an important question: Was the hairy corpse routinely shipped in one of the coffins labeled DO NOT OPEN, which occasionally contained corpses stuffed with heroin, or was the imaginative scheme devised to slip the hairy man through, the very origin of the method then used to smuggle heroin?

When Hansen found himself with a problem he had to solve, namely how to get his specimen to the United States, an obvious solution must have come to mind: What better place to hide a corpse than in a container that is normally used to carry corpses—a coffin. So, since each day a number of coffins left Da Nang airport, where Hansen's unit was based, it was really difficult *not to think* of that solution. The only problem was to ensure the complicity of one or more of the men in charge of placing those more mangled bodies into the coffins. Since some of the coffins were not to be opened for *any reason*, the clandestine operation was almost guaranteed to go undetected. Now, to think of using that method to ship out packages of drugs needed a little more scheming. One may then wonder whether it was not the transfer of the hairy corpse that gave to others the idea of applying the same technique to a much more lucrative contraband.

If that happened, the simple trick, born perhaps from Hansen's imagination and used for a rather innocent purpose, might have turned out to be the pilot project of a large-scale criminal activity. In that case, the introduction of the pongoid man into the USA would no longer be just another incident, or some error committed by an over-greedy or helpful accomplice of the drug trade. It would have been the very beginning of an operation that over the years must have yielded billions of dollars to its initiators. Protecting themselves against any revelation by the ex-aviator as to the mechanism of the operation would have been even more crucial. Given the *modus operandi* of people in that line of business, one can understand Hansen's panic upon publication of my scientific note; my revelations exposed him to having to explain the origin of his specimen and its importation into the United States.

It's not my business to determine what role Captain Hansen of the 343rd Fighter Group might have played, consciously or not, in the development of the matter of the illegal transport of drugs in coffins; I am not the least interested. What seems to me of a great scientific interest is to discover the geographical origin of the pongoid man. This is where the analysis of the situation can provide some precious clues.

A priori, for anyone well informed about the problem of wild men on a global scale, Hansen's specimen could come from almost anywhere on the planet. As far as I am concerned, I believe that the area inhabited by wild men possessing the anatomical characteristics of Neanderthals is much more restricted and excludes North America. The descriptions of Sasquatch and Omah are quite different and the tracks they leave behind could not be produced by Neanderthals.

For anyone not familiar with the subtleties of comparative anatomy, I will call upon a simple psychological argument, based on elementary logic.

If the specimen had really originated in the United States and had been shot there, Hansen would have had nothing to fear from the law. It would have been just an unusual hunting accident, even perhaps a case of self-defense. Nothing that would bother anyone, even less so a captain in the US Air Force. All these slippery maneuvers, expensive precautions, and changes of attitude by Hansen would no longer make any sense or be justified. He would have had a triumphal show across the United States and would have obtained legal authorization to carry his exhibit abroad. Whether through that travelling show or through the sale of the specimen to some scientific institution, Hansen would have made his fortune without any problems.

The hypothesis of a local origin of pongoid man is in all respects completely absurd and doesn't deserve consideration. There can't be any doubt that the hairy corpse came from the Far East, where Hansen had traveled regularly for twenty years. But from which part of that region? Clearly in a country where the American military could circulate freely, either on duty or for R&R.

Up until the day when he first realized how advantageous it would be to claim that he had shot the hairy man in the United States, Hansen stuck to a standard version of his first encounter with it: he had bought it in Hong Kong (one can buy *anything* in Hong Kong).

At first sight, this seems to be a plausible story, especially if one recalls the distribution of Porshnev's troglodytes in Southern China: they would have survived in the Tsinling Chan range, in the provinces of Shensi and Kansu, and in the Hengduan Mountains, in the heart of Yunnan, some 1,600 km (1,000 miles) from Hong Kong.

Hong Kong has been called "the city that opium built." It is in a way the smuggling center of Southeast Asia. Also, because of its location, it is the escape valve of products exported from the People's Republic of China to non-communist countries. So, the sale of a "curiosity" from the interior of Mao's republic would not be anything extraordinary. It is rather the transport of the Thing to the British colony that would have caused problems. It would be difficult to carry some hairy biped over 1,600 km without being noticed. Even more difficult would be to transport the corpse over that distance without it decomposing along the way.

Continuing with this hypothesis, it would not be the transport from South Vietnam to the United States that would have been a problem, but the move from Hong Kong to an airport in Vietnam. Hong Kong, not being part of the military theater, would have no option to take advantage of a transport in some untouchable and sacred coffin.

One is finally led to believe that it must have been from Vietnam that it was easiest to ship the corpse to the United States and that *it was also in Vietnam that it was most likely that the pongoid man had been found and probably killed.*

A number of recent reports have surfaced, confirming the presence of such beings in the Annam hills, beings formerly also known through folklore.

Even since the beginning of this interminable bloody Indochinese war, there have been more people travelling through the jungles than heretofore, with a higher likelihood of coming into contact with these shy creatures.

For example, in 1965, the English journalist William Burchett heard an extremely revealing story directly from one of the heroes of Vietnamese resistance, Tran-Dinh-Minh. The latter was charged, in the days of the French occupation, with setting up bases of the National Liberation Front around the strategic center of Ban-Me-Thuot, in Dak-Lak province, a key location in an area the French called the High Plateaus. In 1949, Tran-Dinh-Minh had been sent to explore, with a few M'Nong natives, the border area of Dak-Lak to find out if the reputedly inaccessible mountains of the Dak-Mil district really prevented access to Cambodia. To his great surprise, he found in the wildest and most arid sector of these mountains a number of prints of bare feet, clearly human. After following these tracks for days, some M'Nong tribesmen made an interesting discovery:

> ... the freshest prints had been made by somebody walking backwards: the heel left the deepest imprint in the sand. We then followed the tracks backwards and they led us to a cave occupied by a very frightened male creature, completely naked except for a miniscule

piece of beaten bark over his genitals.[41] His whole body was covered with a thick black pelt and his hair fell to his shoulders. Huddling in the corner of the cave, the creature seemed overwhelmed by terror, in spite of our efforts to show him that we had no hostile intentions. The M'Nong spoke to him in their language and I tried every dialect that I knew to no avail: we got nothing from him other than sounds similar to those we had heard a little earlier, some kind of bird-like twittering.

It's worth noting that there were found in the cave not only leftovers of a meal in the form of small animal bones but also traces of a hearth, a few sharp cutting stones, and even a kind of bark sleeping bag tied together with vines. That leaves one to suppose that in some regions, relic Neanderthals, if that's what they were, possessed some rudiments of tool making. Of course, nothing proves that the hairy creature was actually responsible for these various signs of tool-making culture.

Minh added that the M'Nong were not as surprised as he was:

> They told me that their brothers of the Dak-Mil district knew of the existence of these strange mountain creatures; they sometimes come upon their tracks when they are following that of a tiger or some wounded wild animal and even to get a glimpse of these "hairy creatures" walking hand-in-hand in the forest. They have seen some of them felling palm trees using the edge of their hand…

Minh and company decided to bring the hairy man, suitably tied up, to their base to try to understand his language and to find out from him some useful information about the area. They found that they had great difficulty in feeding their prisoner along the way. He absolutely refused any cooked food, whether it was rice or roasted monkey (which seems to suggest that he might not have been the author of the fires that had been lit in the cave). Finally, he accepted to eat the leaves of a kind of palm tree, which locals claimed he fed on, as well as raw meat.

Once at the base, they came to the conclusion that nobody was able to translate the hairy wild man's twitter. Since the palm tress that he fed on did not grow in the area, they decided to take him back to his mountains. Unfortunately, he died along the way and was buried.

---

41   According to the M'Nong whom Burchett later questioned, these creatures were "always naked, males as well as females," which led the English journalist to comment: "I could not verify this particular point with Minh, but it is possible that he might have added this detail—a typical example of Vietnamese tactfulness—so that I would not believe that he had traveled for weeks in the company of a naked human being. It is not impossible that he may himself have made the bark modesty shield he had spoken to me about."

Something is certain: in the Dak-Mil district, everyone knew about the existence of these creatures described as "the wildest and shyest that can be."

I also found another story, even more recent, which took place relatively nearby—within 500 km (300 miles)—somewhat less well documented, but which refers to a contact between American soldiers and what might just be one of these elusive creatures.

Writing in the November 1, 1966, New York *World Journal Tribune*, war correspondent Jim G. Lucas described his visit with the South Vietnamese soldiers posted near the demilitarized zone: "The world is somewhat unreal here along the DMZ. The men wander about in the jungle. Once they saw a tiger. They shot at it but missed. Other marines report that they killed a huge ape."

The only known anthropoid apes from that region are gibbons, the smallest kind; no one would normally describe them as "huge." Why not think of pongoid men?

No date is mentioned for that incident; it could well have happened months before the journalist's visit. Who knows? Perhaps this huge ape might be that very same hairy man I examined in its frozen coffin in Minnesota in 1968 and which I then studied for many years. Perhaps Hansen, who retired from the Air Force in November 1965, happened to visit the base of those marines and saw their strange hunting trophy, which gave him an idea…

Anyway, if such an incident could have happened once in that region, it might have happened again. The guerilla war disturbed the most secret hideouts of Indochina's hairy wild men. The raining bombs and rockets, flamethrowers, grenades, and machine guns must have terrorized those innocent savages. Perhaps it happened that while fleeing some terrors they might have run into the arms of equally hostile and ferocious invaders. Perhaps sometimes they approached, carefully curious, the encampments of those terrible green men with polished helmets, falling from the sky suspended from enormous white mushrooms, and spitting fire from their monstrous black sex organs. And sometimes perhaps, in spite of their extreme care, they were seen by these undesirable Martians and either killed or captured.

The whole crazy business of the frozen man could perhaps have started with some accident involving idle soldiers.

It seems to me, looking at that news item, that we have reached the origin of that whole business, not that it happened exactly like that, but that it could have followed such a logical and coherent pattern, given what we know. Whatever variations one might reasonably imagine, one always returns to a solid sequence of events without gaps or dark spots, which leaves no questions unanswered. If doubters who see this as self-deception or error wish to propose a more likely explanation, I am ready to listen to them.

So listen up, here is the story, if you haven't already put it together yourself.

One day, in the 1960s, in the midst of the Vietnamese jungle, a bored GI shoots down an unfortunate Neanderthal whom he mistakes for some kind of gorilla. A rather eccentric flyer hears about it, dreams of turning this hairy monster into a sensational fairground attraction, and buys it. In order to ship it to the United States, the corpse is of course put into a coffin, and more specifically, to avoid attracting attention, in a coffin supposed to hold the remains of a soldier who died in action. Unfortunately, this subterfuge is, or will soon be, what's used to introduce massive shipments of heroin into the United States. Another complication is that the drug shipments originate in the Golden Triangle, where the traffic is organized by the military leaders of US allies, with the blessing, even the complicity, some would claim, of the US State Department or US intelligence services.

Back in civilian life, the owner of the hairy and now properly frozen corpse begins to exhibit it in commercial fairs. However, held to absolute secrecy as to the manner in which the corpse was brought in from Vietnam (there is no joking with the Mafia), he has taken the precaution of protecting himself by having a replica made, whose fabrication will easily be authenticated. Alas, this precaution constrains him, just so he is not convicted of fraud, to leave a veil of doubt as to the authenticity of the specimen, an ambiguous attitude that has a negative impact on the success of the exhibit. He thus dreams of being able, some day, to present his hairy specimen as the real "Abominable Snowman" of the Himalayas. But is that creature actually related to his specimen?

Hansen learns that similar creatures have been seen in the Pacific Northwest and begins to wonder whether he might not be able to claim his Southeast Asian creature to be one of them. He would then no longer have to explain the illegal importation of his corpse or worry about the legal implications. But might there not be some detectable difference between his Vietnamese ape-man and those of Nepal or North America? Only a specialist might be able to tell. So, just to be sure, Hansen decides to consult the most famous one in the United States, a writer-naturalist. He attracts him by having an accomplice call him with a particularly suggestive description of his fairground attraction. The expert swallows the bait and even brings along a colleague from France. The two zoologists soon recognize the authenticity of the corpse and its incredible scientific value. However, against all expectations, the first expert alerts the FBI to have the specimen confiscated, and the second approaches the *Smithsonian Institution* with the same intent and publishes a scientific note about it in an overseas journal.

Justifiably, Hansen's financial backers and/or helpers in smuggling the specimen are worried and angry. They threaten him with the worst and order him to put precautionary measures into effect. The hairy man is then withdrawn from the exhibit, thawed and modified slightly, and then most carefully refrozen and put forward as a fake. The artisans who contributed to the manufacture of the dummy come forward to confirm this "revelation." Most of the scientific or administrative organizations are duped by this ruse and find themselves relieved not to have to be concerned with a scientific discovery that throws doubt on many accepted dogmas.

Of course, the fairground show loses much of its interest in the wake of this demystification and Science loses an extremely rare specimen. Only a few persevere in their search for the truth. They now face overwhelming difficulties, or at least a blank wall of inertia. The problem is that the truth would expose the stratagem of using the coffins of heroes to smuggle heroin. There are too many powerful interests at both ends of this affair, from reasons of state to sordid financial transactions, including the prestige of the armed forces, whose members actively participated in the operation.

So it is that when there occurs a unique opportunity for officials to verify the authenticity of the specimen, at the crossing of the Canadian border, word from above firmly holds back the arm of the Law.

It was a close call! Wishing to make his exhibit more attractive and at the same time hoping to find a permanent shelter from further investigations, Hansen suddenly changes tack. He proclaims widely the authenticity of his hairy man and goes as far as claiming that he shot it himself during a hunting trip within the United States. He says that he is ready to submit the specimen to scientific scrutiny, as long as he will be granted impunity from any infractions he might have committed in this affair.

This clever proposal doesn't trigger any official response. Twice, through a reputed scientific institute, the overseas zoologist begs an advisor of the president of the United States to take Hansen up on his offer or at least to do something to get hold of the specimen, whose study is of major importance for all of mankind. His requests are completely ignored by the White House.

Finally, one day, a cease-fire is signed in Vietnam and, at least in appearance, the CIA decides no longer to ignore the issue. Federal agents suddenly decide to blow the cover on the macabre heroin pipeline that brought the specimen into the US. But —Surprise!—the smugglers have been warned, and the sting ends up with the arrest of a pawn whose main offence is to have impersonated a member of the forces and to have used false documents. Because of statements made by some reporters, the business of drug-filled corpses is exposed, but all official agencies immediately try to suppress it, claiming that highly secret investigations are under way, and emphasizing the strictly "civilian" nature of the affair. By that time, however, nobody respects the cease-fire and war continues to rage in Southeast Asia; there are still allies to be treated diplomatically, powerful interests to manage, and money to be made. Darkness persists and daybreak is still far away.

So that's the main thread of this rather complicated but perfectly consistent story. Besides the survival of Neanderthals, which may well be a surprise to those still unaware of it, is there anything incoherent, incomprehensible, illogical, or inexplicable in it?

The facts are all verifiable; most have even been published. Just as Poe solved the mystery of the murder of Marie Roget using only newspaper clippings, I have assembled them in my ivory tower like so many pieces of a puzzle.

Naturally, American friends and correspondents, some of whom are of course members of official agencies, have been extremely helpful in this task, checking on facts and offering clues and contacts. I thank them from the bottom of my heart, for they have helped me in the most disinterested manner, for their love of the truth, sometimes in difficult circumstances. This whole business wanders through the morass of military secrets, inter-service rivalry, diplomatic protocols, the *omerta* of crime syndicates, the lowest denominator of high politics, and an understandable blackout on police investigations.

Now that the puzzle is complete, my more prudent American friends suggest that I should be careful, and they warn me, for my own sake, against publishing my conclusions. There are apparently powers that one should not oppose, powers that do not like people who are curious and nosy, powers that would not hesitate to eliminate anything that would hinder their activities.

Some of these thoughtful friends are members of the police and must know what they are talking about. So, since a useless precaution is better than no precaution at all, many copies of the file that contains the principal documents related to this affair are stored in secure locations in the United States and in other countries. If anything were to happen to my informants or to myself, these documents would be forwarded to the authorities, as well as to a number of American journalists known for their probity.

Let me add that if anyone planned to get rid of me, to punish me for my indiscretion, I would be rather pleased, for that would turn out to be a strong expression of support for my ideas. Getting rid of me violently would prove to the world that I was right. This proof would certainly not be as rigorous as that which I have presented here and which may only convince those who already agree with me, but it would be much more moving, more likely to raise righteous anger, and thus more effective.

That being said, I have no intention to become a premature martyr of science; I still have too many zoological mysteries to solve. As I am not of a melodramatic temperament, I don't really believe that anyone will try to silence me permanently. First of all, because I have already said everything I had to say about this affair, and secondly because it's so much easier to drown out my words. It is not even necessary to act for this to happen; most people are born deaf, mentally speaking, of course. They hear only what already fits in with their well-established beliefs or given ideas. Everything that's new, unusual, or uncommon remains inaudible.

I call this unfortunate condition: incredulousness. I would even think of it as a disease: "incredulititis."

Chapter 9

THE WALL OF INCREDULITY
How to bury an embarrassing corpse

"One would rather write for eight days about how something cannot be than study
for one hour to convince oneself that it might."
— Jacques Boucher de Perthes

"For most people, doubt about one thing is simply
a blind belief in another."
— Georg Christoph Lichtenberg

One of the discoveries that caused the greatest embarrassment to traditional anthropology was made in 1921 in the zinc mines of Broken Hill, in Northern Rhodesia, modern Zambia. It was an extraordinarily bestial-looking human skull, close in its features to the Archanthropians, the group that includes Pithecanthropes, Sinanthropes, and other Atlanthropes, which were eventually lumped under the label *Homo erectus*. This group is supposed to have completely disappeared since the end of the mid-Pleistocene, at least 60,000 years ago. However, that skull was not fossilized at all; it was extraordinarily fresh and accompanied by the remains of numerous modern animals found in eastern Africa. It also showed numerous signs of dental decay, a condition generally considered to be of recent origin.

In their effort to rejuvenate this indecently belated African Pithecanthrope, most authors interpreted it as a Neanderthal having retained some archaic characters and, just to be sure, claimed that it was impossible to date it accurately. Even the most conservative had to admit, grudgingly, that it couldn't be more than twenty to thirty thousand years old, which is not much, even for a Neanderthal. Actually, everything seemed to suggest that the specimen might not be older than thirteen thousand years old.

Needless to say, fantastic rumors circulated about this embarrassing specimen. Its left temple had a small round hole through it, and some claimed that this early Rhodesian had been shot by a firearm! It was also said that its entire body had been found, still with skin and flesh, thanks to the preservative virtues of the zinc salts that permeated the ground where it lay. It was even claimed that the miners who found it decided that it was more precious as an ore than as a "curiosity" and had preserved only the skull and a few bones.

Commenting on these entirely unfounded rumors, Harvard Professor William

Howells sighed, in his book *Mankind So Far*: "Had those rumors been true, there was enough there to trigger an epidemic of suicides among exasperated anthropologists."

I can fully reassure Professor Howells, former president of the American Anthropological Association, about the safety of his colleagues in the face of such an event. This is just about what happened in the case of the frozen man: an intact corpse of a Neanderthal was offered to the scrutiny of anthropologists, right in the United States! To my knowledge, there was no mass suicide—neither within the country nor anywhere else in the world—when the news came out about the disappearance of the specimen. Other than a few rare exceptions, none of them bothered to have a look at the specimen when it was still possible, or to find out about it later from those who had studied it. Not only did no one lift even a baby finger to prevent the destruction of the specimen, but some even helped to make it happen. They competed in their efforts to get rid of this embarrassing corpse, even fresher than that found in Rhodesia, as soon as possible.

In this case, the only biologist who might have been temped to commit suicide was yours truly, because some of his dearest and most respected colleagues, along with a number of lay volunteers, went a long way to drive him to despair.

Without in any way seeking sympathy for my predicament, I wish to bring up a few significant incidents from my personal "struggle for the troglodytes" to show how it is still possible these days to stifle an unorthodox scientific discovery. One of the most common objections put forward against the truth of my story is as follows: given today's development of the mass media, how is it still possible to gag or censure a scientist when he wishes to expose facts arising from his research or to defend a new theory?

There is a very simple way to do this: all that's required is to ridicule him, to dishonor him, or simply to ignore him. People will follow along like sheep. In this type of combat, all blows are permitted.

But why, according to another common objection, would anyone wish to silence a discovery that would benefit everyone since it would contribute to a better understanding of all mankind? To be clear, I am not referring to the excellent financial, legal, diplomatic, or political reasons that, in this case, might account for the pressures and intimidations used to hide the truth. All that has nothing to do with Science. I speak only of what, in a general sense, might justify apparently objective and disinterested efforts at stifling many innovations and scientific revolutions. Those are psychological reasons, which I will return to later.

But first, let's return to the presentation of the facts, and let's consider them in their chronological order from the point of view of the reception given to the description of the pongoid man.

My first task after examining the specimen most attentively was to write a scientific note about it, to submit it to the scrutiny of a few experts, and to publish it as soon as possible. I was of course accused of being hastier than is appropriate in scientific enquiry.

Had it been the matter of a simple skeleton of an unknown hominid, I would have of course acted differently. I would perhaps some day have communicated my discovery to the scientific literature in general terms, and then spent a few years examining the bones, and also their photos, and consulted the most eminent specialists before publishing a detailed description and my conclusions. However, in this case, there was some urgency. The specimen was decomposing and was also likely to become unavailable. I had to act as soon as possible.

But I just could not be satisfied making a few general public statements, which would have been received by smiles and shrugs of disbelief. To convince the scientific community of the importance of the discovery and of the need for immediate action, it was absolutely necessary to describe the specimen in as much detail as possible and to situate it as precisely as possible within the zoological classification: there had to be some idea of what it was all about! However, as subsequent events were to show, this would have taken me at least two years. What would have happened in the meantime to the poorly frozen corpse?

I was caught between two conflicting forces: prudence versus urgency. I had to find some in-between solution, necessarily imperfect: to publish as soon as possible a preliminary note, sketchy and imprecise and possibly containing some errors, to expand upon later with a more careful and detailed study of the specimen's external anatomy and likely classification.

Today, I am being criticized by some for not yet having, after four years, published a monograph on the subject; I shall return to that point. So I am being blamed on the one hand for the hasty nature of my first publication, and on the other for being too slow to publish the final results of my work. In a word, I was both too hasty and too slow! I'd be happy to find out what others would have done in my stead!

Having completed my preliminary note, I looked for other outlets for publication, hoping that it would also appear in the United States, which deserved to partake in the glory of the discovery. I also figured that publication of a preliminary note in an anthropological magazine or other reputed scientific journal would weigh heavily as justification for government intervention in seizing the specimen. Besides, as it was also important to alert public opinion to the issue, I hoped that some of my photos, together with a brief description, could be published in a mass circulation magazine, such as *Life, Look* or *National Geographic*. It seemed to me that publishing only in a more specialized magazine, such as *Argosy*, as illustrations for Sanderson's article, was likely to lend the whole affair a sensational and shallow aura.

So, although I was at that time just about to leave for an expedition in Central America, I had to stay a little longer in the United States.

At this point a tragic incident occurred in my private life. I would never have mentioned it here if this incident had not been exploited to hinder my work. On January 1, 1969, I received a call from Europe telling me that my daughter Anita, about twenty years of age and whom I had left in good health three months before, had been struck by a particularly noxious form of leukemia and was given only two or three weeks to live. One can imagine the impact of such news and how powerless and torn I felt and the very moment when I had made a major discovery that kept me on the other side of the Atlantic. It was apparently not enough that this incident affected my private life. Some even claimed, in order to discredit my conclusion as to the nature of the frozen man, that the unfortunate death of my daughter had affected my mind. A shining example of the serene, polite, and lofty climate surrounding scientific debate! Anyway, here as elsewhere, the show must go on.

As for me, this was the time when I was seeking the support of influential colleagues as well as publicity in reputable publications. I busied myself all across New England, from Maine to Washington, D.C. The winter was severe, and I remember that my friend Jack Ullrich and I found ourselves during one of our trips stuck and buried under two meters of snow. I recall walking on a New York sidewalk, on a day cold enough to freeze your nose and your ears, with under my arm an enormous portfolio filled with enlargements of my photos and the first interpretations of the anatomy of the specimen. Hating as I did to beg for anything, I nevertheless found myself acting like a travelling salesman for Science, going from door to door to sell my merchandise.

Often, as I crossed paths with pedestrians in this city where everything seems possible and nothing is surprising any more, I would find myself thinking: "If only they knew that I am carrying with me photos of a prehistoric man…They will surely be shocked and surprised when they see them and realize…" I was still harboring illusions!

For example, one day I had a meeting at the Time-Life Building with Al Rosenfeld, one of *Life* magazine's scientific correspondents, whom I had met at a cocktail party when I arrived in New York. A team from the science department of the magazine hovered around my photos, commenting and asking questions. At that point, the chief science editor came by, a woman who was the very prototype of these American career women whose sophisticated elegance usually can't quite hide their physical plainness. She had crossed the room like a hurricane, picked one of my photos by the corner and cast a quick glance at it over her horn-rimmed glasses, then asked a brief question and put the photo back on the table with an air of complete disgust, mumbling: "Oh! That kind of stuff!"

Needless to say, not all of *Life*'s science reporters were like that, otherwise that marvelous magazine, so well documented and illustrated, would never had been what it was; I must say that I cried over its demise, which I saw as a sinister symptom of our times, an abasement of values and the failure of quality.

Al Rosenfeld and his team were keenly interested in publishing my photos, but the editors insisted in obtaining a second opinion from an American anthropologist. At that time, Hansen still absolutely refused to allow examination of the specimen. So, I suggested that they call on Dr. Carlton S. Coon, who had studied my photos and had convinced himself of the authenticity of the specimen. But what was asked was the opinion of a specialist who had examined the actual corpse, which at that time was still impossible. Finally, I asked: "Do you believe that I am not myself qualified to recognize whether such an object is real or false?"

"No," they replied, "everyone acknowledges your scientific credentials and recognizes the care with which you address even the most controversial zoological problems. However, it is quite possible that you might have been mistaken, or taken advantage of…"

"Of course," I said, "but the other expert might also make a mistake, one way or the other. That will not guarantee the authenticity, or lack thereof, of the specimen. I fully agree that merely on the basis of my photos, it might be difficult to be sure, at least at first sight. However, I have personally studied the object at length, and I found some anatomical details that a hoaxer could not possibly have thought of creating, so that I am taking full responsibility for my conclusion. If I had any doubts whatsoever about the authenticity of the specimen, I would not so lightly risk my scientific reputation. What would I look like if the autopsy of this specimen revealed it to be a dummy!"

"Of course," they said. "But you must understand that in any case we cannot rely uniquely upon your statements. All this means a lot of money to us…"

I was briefly taken aback. I had never thought about that; I must confess that I have never been very interested by money. Publishing rights for my photos in *Life* would be on the order of $35,000, according to the scale appropriate to such exceptional documents. That was a considerable sum of money, which would have certainly helped me in my research activities, which were usually carried out on a shoestring. Even if I had been tempted by greed, it would have been most gauche to take advantage of a situation where a mistake or a hoax would eventually have been discovered, thus ending my research career forever. Who would have taken me seriously after that?

So in order to convince the editors of *Life*, I ended up saying: "Listen. My interest in this business is strictly scientific. If you think that I am motivated by some financial interest, I propose the following. On the contract for the sale of these pictures, *Life* will only pay me for the cost of my stay in the United States ($2,000 to $3,000); the rest can go directly to the World Wildlife Fund."

This proposal had an effect completely different from what I expected. All of a sudden, I was seen with mistrust and suspicion. In the United States, at least, everything revolves around money. Someone like me who doesn't have much money but happily spends thousands of dollars on an institution devoted to saving the last Asian rhinos and orangutans, to stop the hunting of tigers and the massacre of seals, and to abolish the trade of exotic animals and the torturing of monkeys in laboratories, must be out of his mind.

That particular episode is characteristic of my personal struggle to communicate our discovery in the mass media. These efforts brought a number of humiliations and disappointments, which I endured with secret pleasure at the thought of the imminent revenge that I would eventually enjoy. Quite naturally, I had met more support and understanding amongst my scientific colleagues than among the public. So, once the affair was in the hands of the Smithsonian, I expected to leave the country without worries. Besides, I had no wish to face the assault from the newspapers, radio, and television once the astounding news would be officially announced. I much preferred to rub shoulders with a different fauna, the real fauna—watching the Yapok on the shores of Mexican rivers, listening to the roars of howler monkeys in Guatemala, looking for the last surviving giant mergansers of Lake Atitlan, or diving in the magic coral reefs of the cayos of Belize.

It was while I was in Mexico, at the Belgian embassy, that I heard the first ripples of the explosion I had triggered. There was indeed a veritable wildfire in the press, a reaction that greatly surprised me after all the rejections and the extreme diffidence I had experienced. In the meantime, of course, my brief scientific report had been published by the Institut Royal des Sciences Naturelles de Belgique—Oh! The glory of the printed word!—and the discovery of the apparent survival of Neanderthals was first page news in all the Belgian newspapers. The shock waves of this revelation would soon propagate through the press of the whole world.

But what condition were these ripples in when they reached the public? It was clear that the information, twisted through the prism of some sensational newspapers or presented with irony by other more serious media, would have seemed to most people just another one of those bizarre items occasionally appearing in the press. Such items sometimes hide a grain of interesting truth but are most often greeted with a doubting smile by a public more skeptical than one might imagine when faced with *unusual* events. We find that today reports of flying saucers and extraterrestrial visitors are no longer considered any more unusual than airplane hijackings or celebrity divorces, but that encounters with surviving Neanderthals still are. It's a matter of fashion, in a society more interested in the future than in the past.

When an unusual fact is reported, it seems the mind rebels, immediately looking for weak points and inner contradictions or impossible claims in the story, and desperately seeks some prosaic explanation. Here's an example.

Essentially, the press had announced that the fresh corpse of what looked like a Neanderthal had recently been exhibited in fairgrounds and had finally been located by scientists in Rollingstone, Minnesota, at the house of its exhibitor, Frank D.

Hansen. Needless to say that it's in Minnesota that this news caused the greatest stir, as well as much surprise and doubt, since no one there was aware of it.

Just to be sure, a reporter from the *Minneapolis Star*, Edward E. Schaefer, had the brilliant idea of writing to me, on March 17, care of the Institute in Brussels, seeking further information. I can't resist quoting a few lines of his humorous letter:

> Our enquiry has shown that the town [of Rollingstone] does not have a fairground, that none of its 392 inhabitants has ever heard of Frank D. Hansen, and that there is no Neanderthal man around, except perhaps on Saturday night after the local bar closes. Is this news based on fact? Is there some explanation for the mention of Rollingstone, Minnesota?

It's because of his particularly strong professional ethics that Edward Shaefer took the trouble of going to the source of the information, but it's quite sure that most newspapermen would have been happy, given these negative results, to conclude that the whole thing was a hoax and to throw the news in the wastepaper basket. The best one can say about said enquiry is that it had been neither thorough nor particularly clever. Nowhere was it said that the scientific examination of the specimen had been carried out on a fairground. In fact, Hansen had actually been one of the 392 inhabitants of Rollingstone for the past eight years; his ranch was located in Crestview, Acresfields, where he could easily be reached by phone (which hundreds of journalists were soon to do).

What conclusion is one to draw except that the journalist in charge of that inquiry had been so convinced *a priori* that the whole thing was a baseless rumor that rather than checking with the local phone company or post office, he had been happy to visit the local bar before closing time. And in doing so he missed a fantastic scoop for his paper, for at that time, Hansen had not yet taken away the specimen and a determined reporter would have been able to hit the jackpot, or at least find part of the truth.

Overall, the press initially limited its role to publishing extracts of my preliminary note, together with various comments. Of course the news was not received everywhere with pure enthusiasm, as in Belgium. Even though the reception at home was most favorable, I could not help deploring the fact that factors linked to national pride would play a role in a purely scientific matter. Elsewhere, there were comments, reservations, and objections; a more normal response, it seemed to me. In London, the *New Scientist* reported that British anthropologists consulted about the discovery had expressed interest as well as prudence, and were awaiting a more detailed study to make up their mind. However, just to be clever, that magazine had used the headline "Cool reception for Neanderthal Man," which did not reflect the tenor of the article.

Although those were harmless scratches, they nevertheless indicated a deep climate of opposition and deprecation. There was much worse to come: frontal attacks, anathemas without trial, unconditional hostility.

For example, a German-Swiss daily, *Blick*, wishing to complement its short report on the discovery, thought of asking the views of one of the world most respected experts on the subject, Dr. J. Biegert, director of Zurich University Institute for Anthropology and also chief editor of *Folia Primatologica*, a renowned primatology journal. Here's what this expert said: "The thesis according to which Neanderthal creatures might still be living among us is total nonsense (*absoluter Unsinn*), whatever supposed 'proofs' might be put forward. Today, on the Earth, there lives only one species of hominids, modern *Homo sapiens*."

This answer is the most striking demonstration of the dogmatic belief in the complete extinction of the Neanderthals, a dogma against which foundered the extensive theoretical studies of Porshnev and his collaborators as well as my own discovery based on material remains. One is reminded of a papal bull: "Whatever proofs might be put forward..."—such proofs being immediately labeled as "supposed"—nothing can change or alter what appears to be taken as revealed truth, an article of faith suffering no debate.

It is as if when LeVerrier calculated that an unknown planet existed beyond the orbit of Uranus, which Galle had then spotted with his telescope, one had stated that: "The thesis according to which an eighth planet might circle the sun is a total nonsense, whatever 'proofs' might be put forward. There are only seven planets in our solar system: Mercury, Venus, Earth, Mars, Jupiter, Saturn, and Uranus." Thus nothing new can be discovered because known entities have been catalogued once and for all, and their number is deemed enough and satisfactory.

Are we dreaming? Have we suddenly, by the flick of a magic wand, been taken back to the dark days of the Middle Ages, or to the heart of a primitive tribe attached to a naïve worldview? Not at all. This opinion was expressed in Switzerland in 1969. Moreover, it was not put forward by some ignorant peasant from the back hills, but in the city of Zurich, from the office of the director of the local university's Anthropology Institute.

It's too much to believe. Surely, Professor Biegert, the author of excellent primatological studies, could never have said such things! For sure, evil reporters misquoted him. Nevertheless, a great many German speakers and others elsewhere probably took his views as gospel truth. With one blow, pongoid man was transformed into an impossibility, or an immaterial creature and perhaps even into a demon...its likely historical fate.

However, a number of people, some of them noteworthy, continued to follow the tracks of this ghost. King Leopold III of Belgium was keenly interested in this discovery, not merely because one of his most loyal subjects was involved, or out of concern for the prestige of his Royal Institute of Natural Sciences, but because he had always had a passionate interest in the most primitive people of the Earth. His Majesty had discussed the matter with Prince Peter of Greece, a renowned ethnographer, who had long entertained a close interest in the Himalayan Abominable Snowman. On his way through Hong Kong, where Hansen's specimen was said to have been procured, the Greek prince had asked a friend, Harold W. Lee, to enquire locally.

On April 30, 1969, that gentleman reported to King Leopold that he had been "unable to find the least bit of information suggesting that such a specimen had been acquired in Hong Kong." He added that the story smelled of a hoax, an opinion he bolstered with a newspaper clipping wherein Mr. J.D. McGregor, associate director for Commerce and Industry, was quoted as saying: "It must have been some hairy dog caught in the hills of Kowloon."

After the peremptory pontifications came the ridicule. Perhaps ridicule can't kill, but it leaves wounds that never heal. I am sure King Leopold must have been pleased to learn that a researcher from his Institute was suspected of having mistaken the corpse of a stray dog for that of an unknown hominid.

Things went from bad to worse after the Smithsonian's press release, which revealed that a replica of the specimen had been fabricated in Hollywood and that one might "reasonably" conclude that there had never been anything else.

By then, there was no need for baseless ex-cathedra pronouncements or attempts at mockery: there was a real bone to chew. And what a bone! An enormous rubber dummy studded with the hair of a bear and sprayed with red paint to enhance the sense of horror. Would-be detractors pounced upon it like a horde of hungry Kowloon stray dogs. What a relief! No more embarrassing "proof" to abuse or ridicule, or even to melt away. No more awkward Neanderthal! It would now be easy to get rid of this pongoid man dummy and its impertinent godfather Heuvelmans, that scumbag responsible for the whole affair.

So without even bothering to learn more, and merely on the basis of an article entitled "The Iceman Melteth" published in the journal *Scientific Research* that suggested that the specimen might well be nothing but an artificial fabrication of rubber and hair, a committee of anthropologists, paleontologists, and zoologists hastily gathered in the USA to formally request that the International Convention of Zoological Nomenclature draw on its authority to cancel, on the basis of article 1 of the Code, the scientific binomial *Homo pongoides* HEUVELMANS 1969. That article forbids attribution of a zoological name to an entity which is not "an animal reputed to exist in nature."

For the benefit of historians of Science, and to discourage prominent scientific researchers from attracting ridicule by making hasty pronouncements about facts they know nothing about, here's the membership of that committee: Ian Tattersall, Peter C. Ettel, Elwyn S. Simons, David Pilbeam, K.S. Thomson, J.H. Ostrom, all from the Peabody Museum of Natural History of Yale University; G.E.Hutchinson, of Yale's zoology department; and U.M Cowgill of the Biology Department of the University of Pittsburgh.

Elwyn Simons, a paleontologist famous for his work on the oldest known hominids, but apparently less favorable to the more recent types, even asked John Napier if he might add his name to the above list, as a co-author of the request. However, Napier, who was better informed on most of the facts in the matter, categorically refused. He even reminded his colleague that he was completely unjustified in expressing such a request, since there was no proof that the pongoid man did not exist, and that he was basing his views on a mere opinion (which is also forbidden by the Zoological Nomenclature Code).

Unfortunately, and precisely because of Napier's ambiguous declarations, the damage was done. The worm of disbelief was in the apple, which it would soon devour completely.

Between the statement that one could "reasonably" believe the specimen to be a fake, and the belief that the fabrication had been formally established, there was but one step, which most people were happy to take. News appeared in print that an autopsy of the corpse, conducted by specialists from the Smithsonian Institution, had revealed the corpse to be a rubber dummy. Of course, this implied, when it was not said more explicitly, that I was a forger, a hoaxer, or a dupe.

In vain did I try, through the press, to set the facts straight and to present the rigorous reasoning that had convinced me. But nobody allowed me to respond, except *Personality*, a South African magazine. In the USSR, the magazine *Prostor* suddenly decided not to print an article that had already been accepted for publication. I was silenced. Any Tom, Dick, and Harry or other ignorant commentator could express his views on the matter, except those most closely involved, the only ones who were well informed. Even Sanderson gave up, at least temporarily; he published a note in his own newsletter entitled "Ends 'Bozo,' We Think." That was it. No turning back!

Now that I was down on my knees and unable to defend myself, it was the perfect time to finish me off. Having excommunicated me for challenging the Dogma, having depicted me as insane, and ridiculing my conclusions to silence me, I had to be kept down at all cost. Defamation was the foolproof way to achieve this.

Please don't imagine that I have a persecution complex. I can produce evidence for everything I write. Shortly after my departure from the United States, an underhanded defamation campaign was launched after me. Its first significant manifestation was the absence of New Year's wishes from the many American friends who had faithfully sent them in the past. I was stunned. I did not understand. It's only a few years later that I understood, when a true friend sent me copies of some libelous letters.

Apparently, I had dishonored myself by publishing, in spite of having given my word (?), a scientific report on Hansen's specimen. Because I had published confidential information, my press card was revoked (…difficult to do: since I was not a journalist, I didn't have one). Furthermore, again for that reason, I had apparently been "banished from the scientific community," a ceremony presumably related to a formal military demotion, the likes of which I did not know existed in the world of science.

In the summer of 1970, I was not yet aware of these low blows, but their impact certainly added to the feeling of defeat and debacle that John Napier's "revelations" had ended up creating. For months, I oscillated between impotent fury and despair. If I did not succumb to the latter and take my life, it was not only because I am too much in love with life and nature, but also because such a gesture would have surely been interpreted as an admission of culpability. However, I was so disgusted by the general attitude of the scientific community that I seriously considered banishing myself from it, by giving up all my university diplomas as well as honorary distinctions, returning all those diplomas to those who had granted them, and resigning from all scientific societies I belonged to. I even considered giving up entirely on scientific publication, "seeking on the Earth a remote corner where one has the freedom to be an honest man," among the animals I love, particular the monkeys, which have been clever enough to avoid being trapped by the lure of becoming human.

But I couldn't bring myself to give up in this way when I thought of all those who were counting on me to solve this mystery and bring out the truth about wild hairy men. For I had not only detractors in the scientific community; many first rate zoologists and anthropologists had expressed sympathy and interest and helped me through their knowledge and advice, even though they might sometimes have expressed some criticism of my interpretations and conclusions, as I would of course done myself in their place. That is indeed an example of fruitful scientific skepticism. Since I have revealed and exposed to the sarcasm of future historians the names of those who have shown a hardly scientific, even unscientific, attitude in this affair, it would be unfair not to mention the names of some others who showed a true scientific attitude, through their open-mindedness, their curiosity, and their recognition of the fragility of current theories. Besides those whose interest has already been mentioned, I would like to pay tribute to eminent scientists like Professor Jean Dorst and Rémy Chauvin, as well as Yves Coppens, assistant director of the Musée de l'Homme , and also to these "young turks" of paleontology: Léonard Ginsburg and Philippe Janvier at the Muséum (of Natural History), in Paris, and Claude Guérin in Lyon; my old friend Desmond Morris in England, who did not blink at finding a Hairy Man facing his *Naked Ape*; Professor G. Vandebroek in Belgium; Professor H. Thiele in Germany; Professor Charles E. Reed in the United States; and Professor Teizo Ogawa in Japan.

And how could I forget, even in the depth of my despair, this young primatologist, likely to emerge as a leader in the field, Scott M. Lindbergh, (the son of the famous aviator, now a dedicated conservationist) who generously offered, when this business began, to finance an enquiry in the United States by a firm specializing in private research to get to the bottom of it. And I should not forget to mention the dedication of his wife Alika, who provided the illustrations for this book as well as the meticulous interpretations of my photographs and the full-scale representation of the specimen, a labor of love to which she devoted many hours and her exceptional talent as an artist. How could I forget the generous assistance of faithful friends, such as my English translator, Richard Garnett, the South African archaeologist Harald Prager, the leading writer-scenarist Samivel, and especially this tireless correspondent Aaron Pearl? Too many people had worked to establish the truth, often ready to take risks; I am thinking about this dear friend formerly from the Israeli secret services, who suggested setting up a commando unit to kidnap the specimen, a desperate but realistic solution that only my horror of any form of violence made me reject. Finally, didn't I have some obligation to all my faithful readers, who expected an explanation and had peppered me with questions through all these years? However, could anything actually be done to convince the others, those who obviously didn't want to be convinced?

I was swimming in indecision when the news came of Hansen's arrest at the Canadian border, soon followed by his acknowledgement that the original specimen was perfectly authentic, which gave me a glimmer of hope. Once more I rose to the occasion. I wrote a series of articles clearly explaining the ins and outs of the story and put them in the hands of a press agency for worldwide distribution. Alas, without much success. Only *Manchete*, in Rio de Janeiro, and Mexico's *Excelsior* published the articles, along with some minor European publications. However, in the USSR, Professor Porshnev had finally succeeded in publishing the essential points of the story in *Technika Molodezhi*. But one is never a prophet in one's own land: no French publication was interested. I could only comment on the latest aspects of the question on the radio station Europe 1. Following that, *France-Soir* decided to make a big first page splash about the discovery of pongoid man.

On the flip side, when on November 7, 1969, the director of the Musée de l'Homme in Paris, Dr. Robert Gessain, a specialist on Eskimos—from which he was assumed to be an expert on frozen Neanderthals—was interviewed on Radio Luxembourg, he formally declared that the specimen must be a fake.

The only way I could have countered that statement was with a detailed explanation of this horribly complicated story, as presented in the eighty-page text I had put in the hands of the press agency, and which nobody in France wanted to publish. Rather than exchanging unsupported statements, I preferred to keep silent and to continue to study the photos of the specimen. I was planning to explain my arguments and conclusions in a well- documented scientific monograph.

At best, I had the opportunity to speak about my research on French television on a "Dossiers de l'Ecran" discussion about wolf-children and ape-men, following a series of Tarzan movies.[42] Soon afterwards, the weekly *Noir et Blanc* boldly published a well-researched article by Robert Herrier that covered the main points of the affair.

Almost at the same time, in July 1970, Hansen's pseudo-confession brought the matter back to the limelight, with further confirmation of the specimen's authenticity. Again, preferring to remain silent, I favored a diplomatic approach, trying, as I mentioned earlier, to nudge President Nixon to intervene so as to acquire that priceless anatomical specimen for Science.

Having heard of my attempts, Eric Vogel, a reporter from the *Tribune de Genève*, requested a private interview so that I could explain my views at leisure, and promised to publish them faithfully, which he did admirably. A full page of that reputed Swiss daily was devoted to the affair, in the issue dated August 22-23, 1970.

There was an immediate reaction from the local scientific establishment. The very next day, Professor H. Gloor, holder of the chair in animal and vegetal genetics at the University of Geneva, sent the paper a letter in which he said that "this business has turned out to be a rather elegant hoax, a kind of Chinese puzzle where there is no lack of artificial hair." And jeeringly, he added, "Before speaking of 'suspense among scientists,' it might be a good idea to check with an institute of anatomy or anthropology."

To which Eric Vogel responded that before speaking of an "elegant hoax," one should also check with the Institut Royal de Science Naturelles de Belgique, which had published my brief scientific report. After having read my report, Professor Gloor had to admit that "it deserved some attention," but that, as a geneticist, he couldn't personally agree with my conclusions. He was also going to seek further information. After consulting a number of researchers, notably from the Catholic University of Louvain and from the Royal Museum of Natural History in Leiden, he communicated his conclusions to the *Tribune de Genève* as follows: "These communications have confirmed my impression that most biologists are convinced that it is a hoax. But I have to admit that, as I originally thought, there is no concrete proof."

In that same issue of October 1, 1970, of the Geneva daily, where again a whole page was devoted to a discussion of pongoid man, Professor Marc R. Sauter, of the Anthropological Institute of the University of Geneva, wrote: "The prevailing impression among specialists is not favorable to that latter hypothesis (meaning the authenticity of the specimen)."

What is one to think of that new scientific approach that ignores concrete proofs and would rather rely on *impressions* from other people, also based most likely on other impressions. One worries, concerned for the future development of human knowledge.

---

42   A year later, I had another opportunity to explain my discovery on television when the O.R.T.F. (Office de Radio et Télévision Française) honored me by inviting me to be their Invité du Dimanche (Sunday Guest) on June 27, 1971.

So, finally, following this display of impressionistic science, the *Tribune de Genève* refused to let me personally refute my detractors, which I would have done by presenting concrete facts and irrefutable arguments.

In October 1970, I had finally completed the first draft of my monograph devoted to the external anatomy of *Homo pongoides,* with some background, based primarily on Soviet and Mongolian research, about its biology, history, and geographical distribution. It was a 300-page text accompanied by 244 references. Before publishing it, I wished to submit it to the views of a number of specialists in various disciplines so as to take into account their criticisms and objections in order to correct, add, or modify the manuscript. Not having the means to produce many copies of this manuscript, which was profusely illustrated but still preliminary, I made three photocopies. I sent one to Moscow, to Professor Porshnev, and one to Bruxelles to Professor Capart, asking them to circulate it among qualified experts. The third copy, I kept to circulate among French scientists. I planned to translate it into English later to circulate it among experts in English-speaking countries. Meanwhile, I was hoping that none of the specialists I consulted would hold one of my three copies too long.

I know very well that in the area of scientific research, everyone is very busy and focused mainly on their own work, but my efforts nevertheless turned out to be very disappointing. At the rate at which the copies circulated, it would take over 80 years! My colleagues' reactions ranged from enthusiastic to absolute refusal even to look at the manuscript. What I never managed to obtain was a list of errors, weak points, and debatable statements. The strongest supporters had nothing to add, and the complete deniers had nothing to say.

Some specialists criticized Professor André Capart for having welcomed in the pages of his Institute's bulletin a preliminary study devoted to what "was manifestly" or "must be," or "could only be a hoax." In response he always told deniers: "If you really think that Heuvelmans made a mistake, or was misled, or has misled us, why don't you say it publicly? Why don't you prove him wrong, why don't you explain your opinion, or your proofs, if you have any? The pages of the Bulletin are open to you as they are to him."

Some replied that they had no time to waste on hoaxes. That was indeed a very poor excuse; the revelation that the Piltdown man was a fake had been a benefit for anthropology, as well as crowning with great glory those who had bothered to demonstrate it.

The deniers preferred to remain silent, not only because they couldn't really have anything to say, but because silence is always more effective in such cases than action. The most hypocritical maneuver was to hold on to the copy of my manuscript for months and months, always deferring a careful reading to a later date, only to finally send it back with a few objections, usually rather trivial. Of course, those objections had already been refuted in the manuscript, which shows that they had probably only leafed through it absent-mindedly, if at all. That a so-called specialist of Neanderthals, for example, wouldn't even be curious enough to peruse a study pretending to prove their survival leaves me rather puzzled.

I would like to single out one very original reaction. During a discussion taking place at a paleontology laboratory on the authenticity of the specimen and the evidence that it brought of the existence today of archaic humans, a young and prominent anthropologist finally exclaimed: "Anyway, what does it matter? That Neanderthals survived to this day does not change anything! The only interesting question at this time is that of the Pre-Sapiens."

I fully agree with that scientist (whose name I shall not mention, as it was reported to me by a third person); the continuing existence of a Neanderthal to this day is completely irrelevant to The Question (that of the origin of man, of course). But to claim that the proof of this continuing existence is of no interest seems to me rather excessive.

Finally, the only worthwhile arguments against publication of my research were:

1. I should have waited for completion of an autopsy of the specimen.

2. One must be prudent in scientific research.

To these I can only answer that only the preliminary publication of a convincing study offered any chance of bringing about the seizure or the sale of the specimen, followed by an autopsy, and that on the other hand two years of study and thinking, including submission of my results to the scrutiny of experts, seemed to me a sign of great prudence on my part.

I could continue *ad nauseam* to list the obstacles of all kinds I met in my efforts to reveal the nature of the frozen man and its mysterious origin. But to what avail? These obstacles always include sullen silence, refusal to discuss, hostility, easy irony, haughty disdain, defamation, accusations of incompetence, inappropriate objections, and advice aimed at dropping the matter. They all added up to absolute incredulity.

Real Science must be skeptical, of course, to the point of doubting of one's own doubts, but it can no more fall into blind incredulity than into gullibility—*a priori* attitudes both. To quote Pierre Bayle: "To believe nothing and to believe everything are extreme positions and both worthless." Unreasoned refusal, loaded with emotion, finds its roots in the Fear of the Unknown, often harking back to early childhood terrors at the discovery of a dangerous world, and to the fears of our remote ancestors setting out for the conquest of new lands where everything was unfamiliar and likely to hide some danger. This fear is also enhanced by a distrust of novelty, based mainly on laziness; it is so much easier to succumb to inertia than to oppose it and it's easier to go with the flow than to resist it. In matters of the mind, this tendency results in a respect for established ideas, for intellectual comfort.

The Fear of the Unknown is overcome only by experience. Once something has been examined, analyzed, tested, and thus known, it is possible to face it. It may still inspire mistrust or provoke flight, but it can no longer cause an uncontrollable terror. That also applies to the Horror of Novelty; experience teaches us that something new is often better than something old and thus preferable. So, Science, even in its most modest form, is the best possible tool for facing the unforeseen dangers of the Unknown, and of the New, which is after all only the Unknown in time. It is so obvious: to adapt to the inconveniences of the unknown, one should try to know it! That is the role of Science! There is irony in the idea that emotionally based incredulity should play a role in scientific enquiry. To do so would have Science denying its own foundation. Isn't it by definition the exploration of the unknown? If it abandons that role, it has no reason to exist.

The systematic prohibition of novelty in science is also a sign of a deep misunderstanding of the philosophy of science, which finally had to admit, following Henri Poincaré, that there is no absolute truth, but that at any one time, it is only the most convenient way of organizing current knowledge.

Truth, if one dares to say it, is provisional. It's not only many-faceted but also constantly changing. This is confirmed by the history of science, which could be also called a history of errors. Copernicus follows Ptolemy, Newton follows Copernicus, Einstein follows Newton, and many since have attempted to blow apart the Einsteinian universe, which they find too static. All have added new bricks to the unstable edifice of Science, or have re-arranged its parts, but all have erred in some way since their system was eventually abandoned in favor of a better one. To imagine that contemporary science is definitive is naïve, silly and pretentious. Such ignorance becomes insolence.

As for me, I know full well that the facts, interpretations, and ideas put forward in this book must include some errors. I can already estimate the uncertainties and weaknesses that exhibit themselves through the differences of opinion between Porshnev and myself. But what we are both absolutely sure of, and which must henceforth be considered a well-established scientific fact, is that *there exists today, on this Earth, another kind of human, wild and hairy*, much like the rest of us, but also essentially different.

That such a statement shocks most anthropologists and primatologists should not be surprising. The entire history of anthropology is an uninterrupted series of capital discoveries that were nearly all received with a horrified denial. So much so that Professor Howell could write that: "…fossil men all seem to have been struck with Tutankhamun's curse, giving rise to interminable disputes and ill-considered arguments." It goes without saying that there can be no scientific area more emotionally loaded than that which touches our own origins. Our ideas regarding the origin of humanity and its races have most often been based on religious or anti-religious dogmas, and on philosophical prejudices and political tendencies rather than on facts. It is thus not surprising that, when even the most minor fact conflicts with established ideas it should be received with suspicion and reluctance.

It is somewhat embarrassing to have to remind the reader of events which should be generally known but which even the experts insist on forgetting, judging by their attitude.

At each significant step in our exploration of the origin of humanity, each time someone delved a little deeper into our roots, the same concert of protestations has been heard, the same collective voice, the same excommunications.

In the days of Cuvier, at the beginning of the 19th century, Science taught that there was no such thing as fossil man, that he had not lived at the same time as the large extinct mammals whose petrified remains had been found.

Many generations of scientists, such as Reisel, Frere, Boué, Buckland, McEnery, Schmerling, Boucher de Perthes, Gaudry, and Lartet, over two centuries, but especially over seventy years, were often the butt of their colleagues' taunts and insults, but accumulated enough pieces of carved flintstone, crafted bones, and remains of human skeletons so that the existence of humans at the same time as the fossil fauna was "officially" recognized around 1860.

The discovery of Neanderthal Man, a more bestial example of humanity, met even more violent opposition, although for a shorter time. Although remains had been dug out as early as 1700 at Canstadt, and that a number of partial skulls had been found at Engis in 1828, at Gibraltar in 1848, in Neandertal in 1856, and at La Naulette in 1865, the existence of the species was not officially recognized until 1886, following the unearthing of skeletons in Spy, in Belgium, in unambiguously Pleistocene strata.

The skullcap of Neanderthal was described by Schaaffhausen as belonging to an ancient barbaric race having preserved, according to Fuhlrott, simian traits, a hypothesis to which Thomas Huxley immediately subscribed. Nevertheless, Rudolf Virchow maintained that the anatomical fragment belonged to a modern human afflicted with rickets and rheumatism. Pruner-Bey, on the other hand, affirmed that its cranial architecture corresponded to that of a modern Irishman, while Blake pretended that it was the skull of a cretinous hydrocephalic, and Mayer that it was that of a Cossack killed in 1814 during the Napoleonic Wars and whose skull had subsequently been deformed by violent blows.

Of course, this dispute started all over again when the remains of what was taken as a real ape-man—a primate halfway between ape and man—were exhumed. A young Dutch army physician, Eugene Dubois, went to Java with the aim of finding there some trace of the "missing link" that Haeckel had imagined and named *Pithecanthropus*. So, in 1894, when Dr. Dubois announced that he had actually found at Trinil a skullcap, a few teeth, and a femur that belonged to that creature, there was a lot of scoffing. Virchow immediately declared that these were merely the remains of some giant gibbon, while P. Minakov, as late as 1923, claimed that the unusual shape of the skull was the result of some mineralization of an ordinary human skull through the pressure of geological strata. One shouldn't be surprised that at the end of his life Dr. Dubois gave some signs of bizarre behavior. Meanwhile, it took the discovery in China of the remains of the very similar *Sinanthropus* to ensure that Dubois fantastic discovery would be taken seriously.

Strangely, although paleoanthropologists are specialists in history, which is said to constantly repeat itself, they don't seem to learn the lessons of the past. The same scoffing and denigration greeted Raymond Dart in 1925 when he published his description of the first Australopithecine, found at Taung in South Africa. At the time, the prominent German paleontologist Othenio Abel declared that it was only a young fossilized gorilla, and others suggested that it was a juvenile of some extinct species of chimpanzee.

Finally, in the 1960s, when Louis Leakey had the bold idea of calling *Homo abilis,* whose skull was found in the Olduvai gorge, a relic older than any of the then known Australopithecines, there was some hesitation about discrediting the man who had been responsible for some of the greatest discoveries of African fossil primates. But there was some backstage whispering that his *Homo abilis* was just another *Australopithecus* and that the nicely rounded form of the skull had been fashioned by a rather forced fitting together of the skull fragments.

In our *Homo pongoides* business, no opportunity was missed to refer to the Piltdown man hoax. It would have been fairer, I think, to cite also the real discoveries of the Neanderthal, Trinil, Taung, Olduvai, and many other locales that were all *a priori* discredited at the time of their discovery.

While in all those cases the passionate and quasi-instinctive denial of the evidence had only served to delay the impact of the discovery and its incorporation into our knowledge, in our case, it may have had much more disastrous consequences, namely the definitive disappearance of the specimen, the holotype of the new form.

Of course, it's not because there are misunderstood geniuses that everyone who is misunderstood is a genius, as failures are wont to repeat to console themselves. It is certainly not because many important discoveries have been resisted and fought against that all unpopular discoveries should be thought important, or have any value whatsoever. And it's not because so many unusual facts have been dismissed that one should believe every extravagant rumor.

In the present affair, however, without even taking into account my own evidence, arguments, and photographs, common sense—and even more so, the spirit of Science—should have led people responsible to assess the problem from a practical point of view, that of efficiency, which is so beloved in the United States.

The choice was then between two positions. That of Napier, who thought that if there was any reasonable possibility that the specimen was false the research should be dropped. And that of Heuvelmans, who continues to hold that if there is the slightest possibility that the specimen is authentic, the only promising approach would be to pursue its study until it was demonstrated to be false.

Following the first choice, there is nothing to be gained if the specimen is false and a lot to lose if it is authentic; by the second choice, there is nothing to lose if it is false and a lot to gain if it is authentic.

Confidentially, haven't you noticed that in this business, those who have strived to procure the specimen and have it analyzed are those who think it is authentic, while those who claim it is a fake quickly lose interest, satisfied with the idea that it was probably so? Doesn't that suggest that the former do not fear being proven wrong because they know that they hold the truth?

Chapter 10

WHAT IT REALLY WAS

A meticulous examination of the specimen, and its identification

"The visible opens our eyes to the invisible."
— Anaxagoras of Cleomenes

"One has to work with what's available."

The entire methodology of the examination of the pongoid man is contained in that statement. Since it was impossible to thaw the corpse, it had to be examined and photographed through its icy tomb. Then, once it had been removed from further study, the only way to proceed further was through the available photographs.

Critics will object that it's not within scientific tradition to draw one's inspiration from folkloric common sense, which should at best be applicable to kitchen problems, Sunday hobbies, or family life. I would answer them with a question: Do paleontologists do anything else when they analyze the meager fragments revealed by their shovels? They do what they can with what they find. Just because zoologists study entire specimens, does it make sense to pretend that paleontology is not a science?

I will venture further. I think that *all* methods of scientific investigation arise precisely from the necessity to work with what's available. Ancient Egyptian surveyors did not have sufficiently long ropes to measure the distance to some point across a river or beyond the horizon. They used what they had, sighting and measuring angles; they invented triangulation, which led to trigonometry. It was impossible to see atoms or their path; Wilson imagined looking at them through a box saturated with water vapor and the cloud chamber was born. It seemed impossible to subject stellar matter to chemical analysis; scientists had to fall back on noticing the absence of certain lines in their spectrum, one of the most spectacular applications of spectroscopy. It was also impossible, notwithstanding some science fiction novels, to descend to the center of the Earth to see what it was like; the study of earthquake waves provided a means to study the Earth's interior safely. In all these cases, one had to work with what was available; one could provide an infinite number of similar examples. The whole history of science is actually an application of that simple maxim.

In our case, the specimen was out of reach behind a wall of glass and a layer of ice. What could be used? What could we hope to do? And how? In other words, what means of studying it were at my disposal? What was I to aim for? And what methods could I use?

The material that I had for study consisted of a series of photos of the specimen and details thereof (about forty personal photos, half in black and white, the rest in color, as well as some taken by others after the partial thawing of the specimen). Some of these photos are particularly clear and revealing to the trained eye, especially those in color, where the diversity of hues allows one to better distinguish details than do simple changes in intensity in the black and white photos. Some other pictures are not as clear, either because of the opacity or lesser transparency of the ice, or because of the abundance of hair, or just because they are only photographs, plane projections of a three dimensional object.[43]

In addition to these photographs, I also had a few sketches taken *in situ* that clarified some details difficult or even impossible to make out in photos.

Finally, I knew the dimensions measured on site, particularly those of the coffin. As I noted before, measurements of the specimen itself, taken vertically through the glass cover, were too inaccurate to be used.

What could I expect from all that?

Actually, what needed to be done with all these documents and data was to reproduce the external appearance of the specimen in three dimensions, *as if it was lying under our eyes on the examination table.* In the absence of the specimen itself, that was the only way of studying its external anatomy and its exact dimensions. This project could be achieved in principle, since I had at my disposal a large number of photos of the "visible" side of the corpse, photos that overlapped and which were taken at a number of different angles. As to the hidden side, since the body was anatomically similar to ourselves, it could be reproduced by extrapolation without risk of major error.

Even then, this was a difficult and tedious task, which took me over a year of daily labor, projecting the photos on a large screen, tracing and interpreting them in actual size, measuring, calculating, approximating, with errors, corrections, verifications…

I know that some would consider all this work a useless waste of time and energy, since someday, the specimen itself might be available for examination. Others may think so but for different reasons; it's possible that the specimen will never be brought to scrutiny and then "there will always remain some doubt as to its authenticity." I have already answered all those objections; it's only through a detailed study of the hairy corpse that one can hope to convince the authorities of the need to confiscate the specimen, and that one will eventually dismiss all suspicion of a hoax. Really, all these objections show that any excuse is valid to stifle such an embarrassing discovery.

I will not dwell on all the details of the task that eventually led me to a complete

---

43   The fact is that we readily fill in the details of a photo when we are familiar with the subject (that is the basis of optical illusions). However when it's a photo of an unfamiliar object or even a known object taken at an unfamiliar angle, we are often puzzled. This is why one preferably uses sketches rather than photographs in scientific publications: line drawings are the favorite illustration in Science; they are in a sense an interpretation of reality.

reconstruction of the specimen; I will only mention the techniques I used, justify my choices, and indicate the steps followed. That will be tedious enough.

One can distinguish two distinct phases in this work. First of all, one had to enhance what was already not quite visible enough in the photos; then, deduce from the visible what was not. In other words, first *decipher*, then *extrapolate*. At the same time, it was important to assign correct dimensions to the whole as well as to the details with respect to the known dimensions of the box, thus to *measure*.

My first step was to decipher the most important photos with the help of the others. One should note that such a process leaves little to the imagination. When done by different competent people, it yields similar results, with small differences in unimportant details. In this respect, one can compare the decipherments I made

A comparison of three interpretations of the photo montage of the specimen (cover) by Heuvelmans (left), by American illustrator John Schoenherr (middle), and by his Soviet colleague P. Avotine (right). A more meticulous interpretation, based on the sum of all available photos, would later resolve the remaining ambiguities.

soon after my initial view, those done from a few photographs by the illustrator John Schoenherr for the American magazine *Argosy,* those done by the Russian artist P. Avotine under Professor Proshnev's guidance for the Soviet magazine *Technika Molodezhi,* and those done by Alika Lindbergh on the basis of an extended scrutiny of all the photographs, which are presented in this book.

The detailed accuracy of these decipherments depends, of course, on the number of photographic documents available to their author as well as on the quality of their individual decipherments.

The task of deciphering is entirely similar to the restoration process used in paleontology, or to the deciphering of texts commonly used in archaeology. In that latter field, there is a clear distinction between *deciphering* a text damaged over time by weather or insects—or where a few letters or simple words, easy to guess, are missing—and *reconstruction* of a text where parts of sentences or even paragraphs are missing. A similar distinction is found in paleontology between the simple *restoration* of a broken fossil where no essential parts are missing (a procedure which includes cleaning, assembling and filling holes), and *reconstruction* of the external appearance of a creature from those remains or of an incomplete skeleton. When half a skull is missing, entirely on one side of the sagittal plane, a direct restoration is possible. However, when it's a mandible that's missing, it is only possible to achieve a more or less precise reconstruction, based on some hypotheses. (One should note however that even the reconstruction of the external appearance of a fossil is not entirely a matter of conjecture, since arguments based on comparative anatomy and even ecology can provide useful guidance.) In summary, while there are various degrees of conjecture in all *reconstructions*, there is a qualitative distinction between any one of those and a *restoration*, which is practically free from conjecture.[44]

Decipherment of the photographic documents first allowed us to see the exact appearance of the face, of the hands, and of a foot of the specimen, as they appear free from their ice matrix.

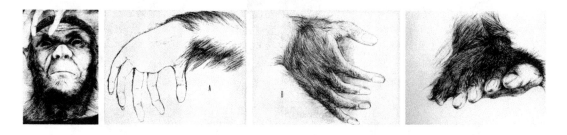

---

44   To emphasize that difference, the plates which illustrate this part of the book have been done using different techniques depending on whether they are simple decipherments of photographs (ink drawings) or reconstructions of the living specimen (charcoal).

The next step was to connect these anatomical fragments while respecting their exact proportions. That required a restoration of the whole of the visible part of the corpse with respect to the coffin, whose inner length and width were known, an operation that had to be done graphically.

Pictures taken at different angles, positioning the contours of the specimen with respect to the sides, corners, and right angles of the coffin, made it possible to build a three-dimensional network (graduated for convenience in 10 cm by 10 cm squares on my drawings) completely encompassing the specimen.

In order to realize this three-dimensional gridding, it was necessary to grid first those surfaces of known dimensions. The most obvious one was the glass cover (220 cm x 90 cm), but there was another one, underneath it and parallel to it, that must have the same dimensions. The body was not merely enrobed in a matrix of ice; it was like a high relief sculpture on a bed of ice formed by partial melting of the block originally deposited in the coffin. Let's call the top of that ice bed *the ice surface level.* Assuming that the coffin is level and that its walls are perpendicular, that surface is necessarily parallel to the top cover as well as to the bottom since it results from the freezing of a liquid.

Based on the laws of perspective, it's then possible to measure on some photographs the distance between the level of ice filler and the top cover (39 cm). It is then possible to grid the whole space between these two surfaces, which contain the visible part of the corpse, emerging from the ice.

In practice, the simplest way to complete the gridding is to work on a photo of the whole taken at an oblique angle from the feet, along the front-to-back axis of the coffin. One then proceeds the same way with two series of partial photos taken perpendicular to that axis, some slightly sideways, the others perpendicularly.

Finally, one can point out precisely the position in space of any point seen from three different angles, and reproduce in space every visible part of the specimen.

Having restored the part of the corpse that emerges out of the ice, one then had to try to reconstruct the part under the ice, which seems a lot more risky. However, after looking at the restored parts, it became obvious *that the corpse had to rest on the bottom of the coffin.* All the contours extrapolated from the upper part converged on the same plane, which was the bottom. The many points of contact between the body and the bottom revealed that it was 17 cm +/- 1 below the ice surface level and thus 56 cm +/- 1 below the glass cover.

This was an extremely important situation, which allowed a confident reconstruction of the invisible part of the body. Imagine for a moment that the hairy man had been struck by *rigor mortis* as it lay on an uneven surface, and that it had been frozen during that rather short period of rigidity (16-24 hours). It would have then been impossible in that case to make measurements such as the front-to-back length of the head, or of the foot. *...text continues on page 164*

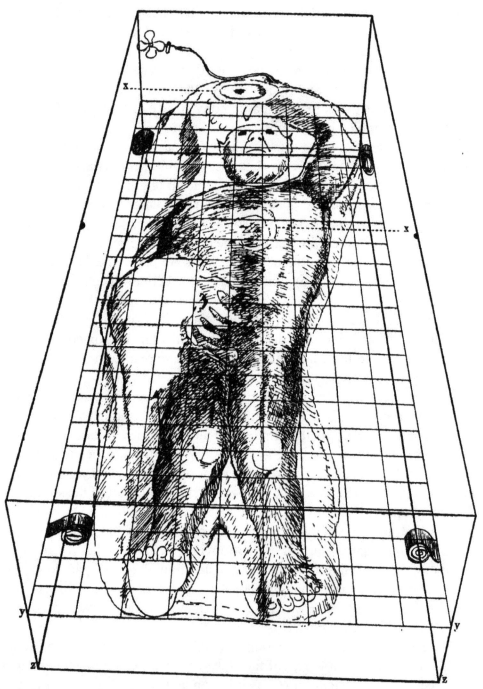

Angled view (45° from the vertical) of the specimen lying in its refrigerated coffin. The surface level of the ice is shown by the gridded network of 10 cm x 10 cm squares. The frosted rings are indicated by the letter x.

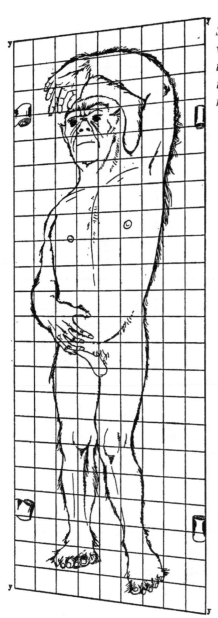

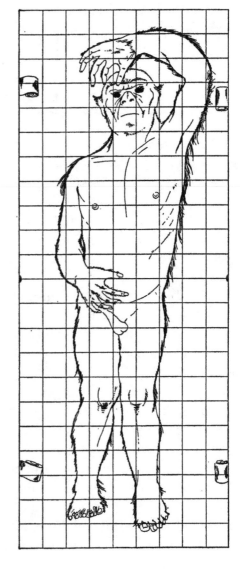

Slightly lateral view (15° E of vertical) of the specimen referred to the grid. Some hair was included to render the anatomy more visible.

Face view of the specimen (vertical view against 10cm x 10 cm horizontal grid).

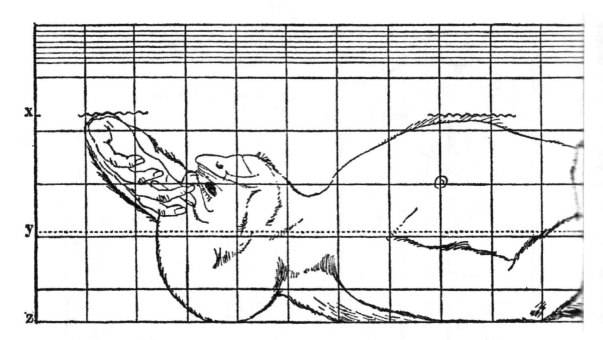

Lateral view of the specimen (against a 10 cm x 10 cm grid) x = upper surface of the original ice block; y = the ice surface level; z = the plane on which the body rests. The four panes of glass making up the cover are shown.

Of course, since the nature of the ice indicated that the corpse had been artificially frozen, the most likely hypothesis was that the body had been deposited while still limp, or after becoming so again, in a freezer, at the bottom of which it had softly settled.

One could even know, at this time, that the freezer had then been filled with water to a height of 40 cm, which just covered the corpse. This is because one notices in the block of ice two surprising structures: an oval frosted ring over the palm and the wrist of the left hand, and a similar crescent in the middle of the chest. Reconstruction of the specimen's volume reveals that these two nearly complete rings are found exactly at the same level, namely 23 cm above the ice surface level, and thus 40 cm above the bottom. I have verified by experiment, freezing a rubber doll, that such structures form around bodies near the water surface. It's easy to understand. As I noted previously, in a freezer, freezing takes place at the periphery, creating a kind of icebox that continuously thickens towards the inside, pushing ahead the bubbles of dissolved air liberated by the freezing of the liquid. These bubbles end up accumulating over the whole surface of the body, but are most concentrated where they are caught between it and the surface sheet of ice, and thus everywhere that the body nearly touches the surface, and around it. That, of course, explains the annular nature of these structures.

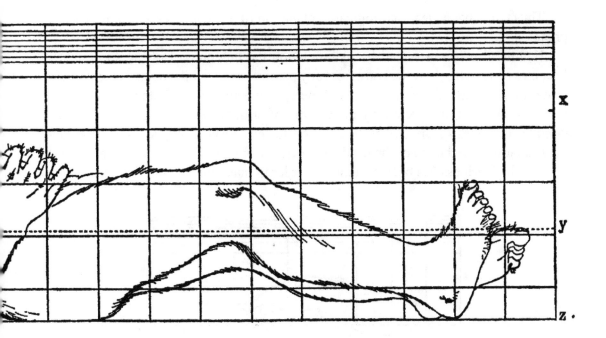

For similar reasons, the increased opacity of the ice around the left elbow and on the side of the left arm suggests that the specimen was originally frozen in a tub no wider than about 75 cm. No analogous conclusion may be drawn regarding the length of that tub, but it's likely that the specimen was frozen in a large domestic chest-freezer, commonly available for sale in the United States. It's common for large families to have them custom made to a desired size. It would have been child's play to find a model with interior dimensions 7 ft long [213 cm], by 2 ft 6 in wide [76 cm].

If it had been created within such a container, the original block of ice, 40 cm thick, would have had a volume of 213 x 76 x 40 cm^3 and would have weighted about 650 kilos [1,460 lbs]. [45]

Using two strong nylon straps, the ice block was deposited into the refrigerated coffin made especially to receive it. Perhaps it is before that operation that the upper part of the ice block was cut down to make the ice covering the corpse as thin as possible and make it more visible, thus greatly reducing the weight of the enormous ice cube. Finally, the filling up of the space surrounding the ice block probably resulted from partial melting during various manipulations.

---

45   Estimating the weight of the specimen at about 125 kilos [281 lbs], it would have been contained in about 525 kilos of ice. If Hansen really poured 20 gallons per day for a week in his freezer, as he claimed, that represents (3.78 liters x 20 x 7) 528.5 liters, which closely corresponds to that weight. This seems to confirm that it is indeed in that kind of freezer that the specimen was first frozen.

Thus, we have a clear and logical explanation of all features seen in the various photos.

Having also reconstructed by extrapolation the unseen part of the specimen, it becomes possible to show the body from any possible angle while maintaining its exact proportions. To avoid distortions due to perspective and to facilitate measurements, the representations are preferably in the form of orthogonal projections on the plans defining the sides of the coffin. In addition to the top view already shown, a side view, a view from the top of the head, and a view from the feet are shown. In all these figures, the hairiness has been sketched in order to make the anatomy clearer.

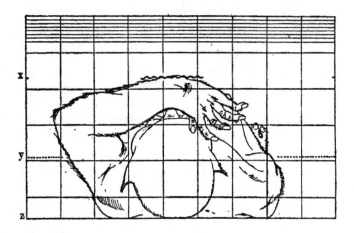

View of the specimen from the head. Y is the ice surface level.
X is the level of the frosty rings.

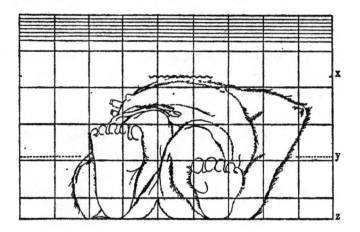

View of the specimen feet first.

Everything is now ready for making measurements, which will of course be repeated on the photos themselves, as control, as well as on graphic projections at various angles.

It's worth noting that the degree of uncertainty and error linked to these representations and measurements, based on multiple photographs, can't be any greater than that linked with similar extrapolations, based on solid fragments, used in paleontology. The techniques we have used here are no less legitimate. The fact that they may be unusual in anthropology is no reason to dispute the validity of these results; that would amount to an opposition to all progress in scientific methodology.

This meticulous reconstruction of the hairy corpse, now seen lying on the examination table, finally yields the measurements shown in the Table of Measurements. (See Appendix B.) All dimensions are given with their error estimates, which are usually 2% to 3% but vary according with the difficulty of finding appropriate anatomical reference points. In any case, the size of the possible error is a lot less than the 10% I had estimated in my preliminary note, a clear sign of progress. The effort was worthwhile.

The creature was without a doubt—I have said it before—a *Homo*; it was thus legitimate, as a mean of comparison, to carry out all the measurements normally made in anthropology on *Homo sapiens.* There was, however, one difficulty.

The height of an individual of our species is normally measured in a standing position; that measurement then enters in the determination of all body indices. But what we have here, as I shall demonstrate, is a Neanderthal, that is a creature with a permanent knee flexure, meaning that it could not achieve a total extension of its legs. It height standing up would be about 4 cm shorter than the sum of all its body segments.

In our frozen specimen, I was inclined to believe that the left lower limb was at its maximum extension and that the right leg was only lightly bent.[46] As I was careful not to prejudge the identity of the specimen in an investigation that aimed precisely at finding what it was, I give everywhere two measures for the specimen's stature and for the total length of the lower limbs, the first corresponding to a standing posture with knees slightly bent (GF), and the second with the lower limbs completely straight (ET). These measurements play an important role in establishing body indices, most of which are calculated using the height.

By calculating classic anthropometric indices established for human races, it is possible to obtain an objective description of the morphological characteristics of the specimen. The various indices so calculated are of course only relative: their only

---

46 For corpses lying on their back, as for people sleeping in that position, relaxation of the muscles tends to place the lower limbs at maximum extension. For a specimen of our species, one then finds the legs extremely straight. The knee is never bent unless the individual is lying on its side or when the sole of the foot lies on the ground.

value is in reference to the norms established for our own species. But their main significance is in pointing out how the specimen differs from our species, especially when the values of the indices fall at the limits or even beyond the range of variation for *Homo sapiens*. Of course, what is abnormal for our species, and might be the subject of teratology, might be perfectly normal for some other species, and be one of its defining characteristics.

To ensure that our representation of the specimen is as complete and as precise as possible, the list of its proportions (seen in the Table of Indices) must be supplemented by a presentation of its purely descriptive characteristics, which are more difficult to quantify. For those, I refer the reader to Chapter 1, where they are presented in detail.

All that was left to do was to identify, if possible, the creature or at least to place it with respect to already known hominid species. In Chapter 3, I already attempted to classify the specimen, to conclude that "there was a strong probability that it was a Neanderthal." However, zoology is not a game of chance. Furthermore, I had reached that conclusion by a process of elimination, a method I cannot disapprove more of as a cryptozoologist, since it assumes that *everything* is known. But it's obvious that the hairy man could well belong to a species completely unknown to paleontologists as well as to neo-zoologists, as we may call those who study only animals currently living.

My conclusion was only preliminary. If I draw upon it again, it is only as a *working hypothesis*. We shall now compare, item by item, the various anatomical characteristics of the specimen with those of Neanderthals to find out how close they are.

But first, a warning. Some will chide me for adopting as a reference the Chapelle aux Saints man, which is often considered as an extreme form and not the prototype of a Neanderthal. As a matter of fact, it is indeed on purpose that I prefer to pick that particular "extreme" example as often as I can.[47] If we are dealing here with a Neanderthal in the broadest sense of the term, we are faced with the contemporary representative of a population that would have continued to evolve and specialize over the past thirty-five to fifty thousand years since its presumed extinction in various parts of the globe.

So, in which direction did the evolution of Neanderthals proceed, particularly with respect to our own species? The analysis of the abundant European paleontological material provides an unequivocal answer.

The flowering and differentiation of the hominids essentially took place during

---

47   The Chapelle aux Saints man is actually not a really extreme form. An extreme always finds another one more extreme, in this case, the Kiik-Koba man (Crimea), which seems even more specialized, and which I will use as a comparison whenever I can. It's unfortunate in this respect that its skull was not found. In any case, the study of its skeleton will guide us in understanding the salient features of Neanderthal feet and hands.

the Pleistocene. During that geological period, our planet, and particularly Europe, was in broad terms subject to four major glaciations. In chronological (as well as alphabetical order) they are called Günz, Mindel, Riss, and Würm, named after tributaries of the Danube River. They were separated by warmer periods called interglacials.

The earliest known Neanderthal, from Steinheim, was found in deposits dating from the Mindel-Riss interglacial. This was a first surprise for those who would think of the Neanderthals as our ancestors: it was a contemporary of an undeniable *Homo sapiens*, the Swanscombe Man (UK) with which he had much in common. Second surprise: there were already some quasi-Sapiens well before that in Europe, during a lull in the Mindel glaciation, as evidenced by some remains found in Vèrtesszöllös, in Hungary.

Neanderthals are first found in France (at La Chaise) during the Riss glaciation and become more common in Europe during the Riss-Würm interglacial. Bone fragments have been found in Germany (Ehringsdorf) and Italy (Saccopastore). However those Neanderthals still presented numerous traits resembling those of Sapiens of that time, some remains of which were found in Fontéchevade, in the Charente area. They were not quite Neanderthals, but rather pre-Neanderthals and thus not classical Neanderthals.

The latter began to show up only at the beginning of the Würm glaciation, as evidenced by the natural cranial mold of Ganovce (Czechoslovakia) and the Gibraltar skulls. They are prominent all along the earlier stages (Würm I and Würm II) of that glaciation, the early Würm. The most famous are from the southwest of France (Le Moustier, La Chapelle aux Saints, La Quina, etc.), Belgium (Engis, Spy, La Naulette), Germany (Neandertal), and Italy (Monte Circeo), but they were also found in Spain (Jativa-Valenza), Croatia (Krapina), and Moravia (Sipka) as well as elsewhere.[48]

It is usually stated that starting with the late Würmian, fossilized remains of Neanderthals are no longer found, which suggests that they would have become extinct before that time. As we shall see in the next chapter, that was certainly not the case.

If one tries to understand something of the problem of the pongoid man, as well as of that of our own origins, one must realize that the evolution of the Neanderthals did not happen, and could not have happened, as it used to be naively believed, starting from a rather ape-like form that would gradually have become more refined

---

48   The Neanderthals also spread to other continents, where they more or less experienced a similar evolution, depending on the climate. In North Africa, they spread from Morocco (Tanger) to Cyrenaica (Haua Fteah) by the end of the Würmian. Some rather unspecialized specimens were found in Asia, in Galilee (Taboun, Djebel-Qafzeh) and perhaps in China (?Ma-pa) during the Riss-Würm, some more specialized at the beginning of the Würmian (Chanidar, in Iraq), and some very specialized in the middle of that glacial age (Techik-Tach in Uzbekistan and mainly Kiik-Koba, in Crimea).

to finally end up with *Homo sapiens.* Irrefutable facts based on comparative anatomy, embryology, and paleontology all converge to show that the opposite happened. As Professor Henri Vallois convincingly showed, Neanderthals evolved from a pre-sapiens form (Vèrtesszöllös, Swanscombe, and Fontéchevade and some others) and gradually evolved into a direction that some have called a blind alley of evolution. Boule and Vallois speak of Neanderthals as a "dead branch…a lateral branch which reaches a very specialized form and then disappears"; Sir Wilfrid LeGros Clark speaks of an "aberrant byway of evolution"; for professor Piveteau, it is "a specialization which one might consider regressive."

Without further pontificating about how presumptuous and anthropocentric it is to consider an evolutionary process that does not lead to our species—"the marvel and glory of the Universe," according to Darwin!—as a failure, a regression, or a blind alley, let's only recall Professor Piveteau's words: "…one fact is clear: *Homo sapiens* does not derive from the Neanderthals that came before him: he was already a member of an independent lineage of pre-sapiens, evolving independently from that of Neanderthals…. As to the classic *Homo neanderthalensis*, which appears at the beginning of the Würmian period, it should be considered as the aberrant end point of an evolutionary lineage, diverging from that of the predecessors o*f Homo sapiens* when they had already reached a stage that *did not differ from today's in any significant characteristic*." (My emphasis.)

This situation has led most anthropologists to distinguish at least two types of Neanderthals, one being the progenitor of the other. The first to appear, and also the closest to *Homo sapiens*, is sometimes called pre-Neanderthal, sometimes a "generalized type"; the second, more recent and specialized, is called "classic" or "extreme."

Through the ages, as *Homo sapiens* remained more or less stable as to its general anatomy, evolving perhaps intellectually, psychically, and socially, Neanderthals—isolated in some regions of Europe and Central Asia and under the selective influence of the harsh climate of the glacial eras—began to evolve and specialize rapidly, showing ever more distinctive anatomical traits. It became more and more "animal-like" and gradually "de-humanized."

If the pre-Neanderthal might have appeared to Sapiens as a rival to eliminate, and perhaps also a likely sexual partner,[49] the extremely specialized Neanderthal was probably seen as mere game, or perhaps a slave, a beast of burden to domesticate.

Should our specimen turn out to be what we suspect it to be, we should expect it

---

49   There were certainly hybridizations between the two prehistoric humans in Palestine. This is clear from the relatively recent (less than 40,000 years) presence of individuals, like those of Skhoul (Mount Carmel), with intermediate characteristics. The supporters of the traditional Darwinian sequence have, of course, regarded them as Neanderthals transforming into Sapiens, but the heterogeneous character of the population in which they are found identify them clearly as hybrids.

to be the final extreme result of the accelerated Neanderthal evolution.

In this respect, I have already answered the objection that claims the specimen is too tall to be a Neanderthal. First of all, the smallness of early men has been grossly exaggerated; they were certainly not quasi-pygmies, as was formerly thought. Carlton Coon pointed out that meticulous measurements showed that the wizened oldster found at La Chapelle aux Saints measured 1.64 m [5 ft 5 in] which is a good centimeter taller than the average Frenchman living in the area when it was dug up. Today, less than 70 years later, the young people of Brive-la-Gaillarde are sometimes as tall as 1.80 m [5 ft 11 in], and the contemporary descendants of a Neanderthal like that of La Chapelle would also most likely be as tall. Especially since over the last 400 centuries, because of their small numbers and isolation in remote areas, they would have been forced to breed locally, which often, in the long term, leads to gigantism.

One has to admit that, other than its height, our specimen corresponds in all its body proportions to the classic Neanderthals: a large head, a very broad face, and limbs of generally human proportions, but with the forearms and the legs relatively short with respect to upper arms and the thighs, relatively short upper limbs compared to the lower, and extremely large, hands and feet.

How about other body features? Is it at all possible that Neanderthals might have been as hairy as chimpanzees?

Not only is it possible, but it's most likely, given the ecological conditions where they lived, or ended up living.

First of all, we should note that mammals are basically hair-covered land vertebrates. While a naked skin is normally found in those that have become fully aquatic (Cetaceans, Sirenians), it is extremely rare among normal species. Besides, in the largest ungulates (Elephants, Rhinos, Hippos), which tend to have a partly aquatic life style, one finds only one of the burrowing cousins, the Aardvark (*Orycteropus afer*) and a burrowing rat (*Heterocephalus glaber*), two related species of bats (*Cheiromeles torquatus and providens*), the Babiroussa warthog (*Babyrousa babyrussa*), and *Homo sapiens*— only 14 nearly naked species among 4,000 species of land mammals. Never has Man been better characterized, morphologically speaking, as when Desmond Morris called it *The Naked Ape*: its lack of hair is what is most original and what most sharply differentiates it from the other primates.

Of all the naked land mammals, Man is the only one that is not specifically tropical. Everything indicates that the extreme-type Neanderthals had profoundly adapted to an intense cold climate, that of the Würmian glaciation. Before that, during the Riss-Würm interglacial period, pre-Neanderthals were only slightly different from the Sapiens and were also more diversified. Some of them, for example those of Saccapastore, near Rome, showed clear signs of adaptation to a warm and humid climate. But after the frost fell, during the Würmian, no more Neanderthals of that type were found in Europe, and neither were any *Homo sapiens*. The only people left were those perfectly protected from the cold, namely the classic Neanderthals.

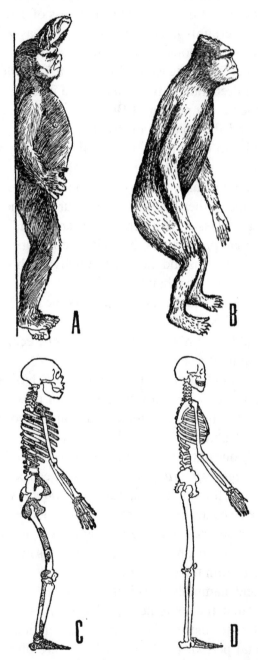

Comparisons of the silhouettes of A: the specimen; B: sketch of the Ksy-gyik (after Khakhlov); C: skeleton of the Chapelle aux Saints Neanderthal; D: skeleton of Homo sapiens.

The various body proportions reflect that specialization: the extremely massive body, the short neck, the head sunken into the shoulders, the arched back, the short limbs, the large hands, the bent knees, and the wide feet. Everything seems to have been oriented in this anatomy to reduce the ratio of surface, where heat is lost, to volume. In this respect, among modern races, Neanderthals were most like the inhabitants of the Polar Regions (Eskimos, Tierra del Fuego natives) or hardy mountain dwellers. Dr. Coon wrote that "People built more or less like Neanderthals can be seen today in the Abruzzi, the Alps and in Bavaria."

The specialized paleanthropians of the Würmian differed from all modern human races by the absence of "canine fossa," small depressions located above the roots of the canine teeth. They are filled up in those hominids by the greater development of the sinuses; it is presumed that this extension of the sinuses helped warm up inhaled air before reaching the lungs and especially the delicate brain. On the other hand, it is likely that the extraordinary development of the brow ridges, which are in fact frontal sinuses, was an arrangement allowing a better protection of the eyes against the cold by placing them deeper in the head.

Isn't it most likely that, other than these subtle anatomical adaptations, the most common and superficial adaptation, the growth of a body pelt, should also have occurred?

The animals whose remains are most frequently found in association with the stone tools and hearths of the classic Neanderthals are the mammoth, the wooly rhino, and the cave bear—characteristic representatives of the arctic fauna. All of these, as we know, were protected from the cold by a thick wooly pelt or very dense fur. Isn't it reasonable to assume that the predators were just as hairy as their prey?

Some have objected that this was not necessarily the case, since the Eskimo and Tierra del Fuego natives, who often live naked in very cold climates, are not very hairy, actually quite the contrary. That may be true, but the temperatures experienced in Europe during the various stages of the Würmian were a lot less mild than those found today in Greenland, Alaska, or Tierra del Fuego. Besides, the fact is that no sure trace of *Homo sapiens*, the "Naked Ape," has been found in Europe in the geological strata dating from the early Würmian (Würm I and II). Although *Homo sapiens* had lived in Europe during the Riss-Würm interglacial, and before that in the Mindel-Riss, he probably had been forced to withdraw towards warmer areas.

It's only when the climate warmed up again around 38,000 years ago during the Laufen interstadial that *Homo sapiens* returned to Europe, pushing aside the Neanderthal hordes and probably massacring most of them. A few thousand years later, colder conditions returned (Würm III), as intense as before, reaching a peak around 20,000 years ago. Logically, one would expect that at that time the Sapiens would have again retreated or perished, and the surviving Neanderthals would have taken over. But that's not what happened. In the meantime, our ancestors had greatly improved their techniques, both tools and weapons, and had also learned how to dress warmly and build cozy houses.[50]

It's very difficult to believe that the extreme Neanderthals were not at least as hairy as our specimen. Their whole history—their dispersion as well as their evolution—is the result of an ever-improving adaptation to cold conditions, which ended up making them very different from the Sapiens. Besides, undisputed Neanderthal remains have been found only in the Palearctic region, namely the temperate and cold Eurasia of today, as well as in North Africa, which was rather cold during the ice ages.[51]

Let's now consider the various parts of the frozen man to see whether they correspond in detail to what we know about the extreme Neanderthals.

Given the flatness of its head and its receding forehead, the elongation of its head from front to back, the extraordinary development of the facial as opposed to the cranial part of its head, and the lack of a chin, it is clear that the specimen has all the features of a classic Neanderthal.

However, although the specimen's head is typically Neanderthal in shape, it is not so by its dimensions.

There is no known Neanderthal skull reaching 21 cm in length.[52] To estimate the corresponding length in a live individual, one has to take into account the fleshy parts. Guerasimov noted that in those individuals where the glabella is most developed, the fleshy parts are thicker than in those where it is less developed. In our species, he added, these parts vary in thickness from 8-12 mm in males and 5-8 mm in females. It seems that in a skull, every bony protuberance is accented by a greater development of the fleshy parts that cover it.

It's thus likely that in Neanderthals, the bumper formed by the brow ridges was probably covered by a fleshy layer between 12 and 18 mm thick among males. As to the famous "bun" (the occipital torus) it was probably covered by the more than 3-4 mm fleshy layer usually found in skulls of our own species. Hence one should

---

50   The appearance of scraped bone artifacts coincides with the disappearance of the Mousterian tools attributed to Neanderthals. One may well wonder whether there wasn't some (perhaps indirect) link between the two. Perhaps Neanderthals were supplanted in Europe in spite of the cold because the Sapiens had learned to make fine needles and to sow warm and waterproof clothes and shelters.

51   It's true that a number of authors have classified Broken-Hill Man, the very similar specimen from Saldanha (South Africa), and the Solo Man (Java) as Paleanthropians, but that's mainly because of their recent age (Upper Pleistocene) and not because of their anatomy. Dr. Carlton S. Coon has clearly demonstrated that the angular, pentagonal shape of their skull, seen from behind, suggests that these various tropical hominids should be classified as Archanthropians. In all Paleanthropians the skull looks perfectly round when seen from the back.

52   Values for the front-to-back (glabella–occiput) distance were given by Morant in 1927 for the skulls of a few adult classical Neanderthals (in millimetres): 207.2 (La Chapelle), 204.2 (La Quina), 200.6 (Spy I), ? 200 (Spy II), 199.0 (Neanderthal), 192.5 (Gibraltar).

add 15 to 22 mm to the skull's length (gabella-occiput distance) to obtain the largest estimated value alive for known classical Neanderthals. Not a single one of those measured would have reached or exceeded 23 cm.[53]

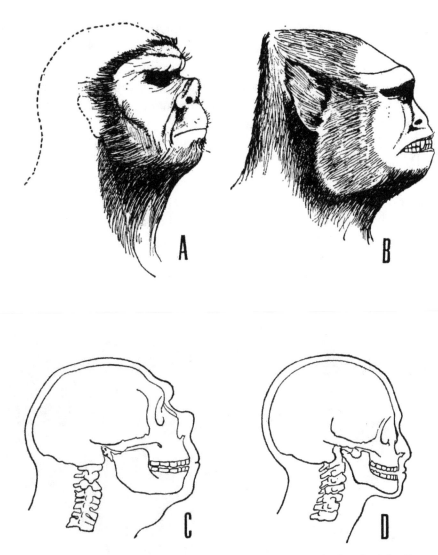

Comparison of profiles of the head of A: The specimen; B: The head of the Ksy-gyik (after Khalkhov); C: The Chapelle aux Saints Neanderthal, with a reconstruction of its profile when alive; D: Caucasoid Homo sapiens.

---

53   For the archanthropian skulls of Ngandong (near the Solo river in Java), one finds a gabella-occiput length of 221 mm, which is the largest value found among hominids. When alive, these Solo Men must have had a 24 cm long head.

However, we found that in our specimen the distance between the tip of the brow ridge and the plane on which the head rests was 27 cm =+/- 1 cm. Assuming, which makes sense, the presence of a thick layer of hair, reaching 1 cm, and a maximum error of another centimeter, we still find a skull length of 25 cm!

Which is enormous. This value of 25 cm +/- 1 is from 9 to 17% higher than the highest such measurement made on a Neanderthal. Other measurements of the head also turn out to be extremely large. Let's take for example, the bi-zygomatic length, the distance between the two cheekbones. In the specimen, it is 185 +/-5 mm, while in the Chapelle aux Saints Man it could not exceed 163 mm (156 mm measured on the skull plus 7 mm for flesh cover). This measurement is thus 10 to 16% higher in the frozen man than in the old man from the ice age.

So, what we have in hand is a specimen with a head that has the shape of a Neanderthal's head, but which is 9 to 17% larger than all known Neanderthals. However, the specimen is also 8 to 12% taller than any of those. All we can conclude is *that his head is proportionally slightly larger than that of the extreme Neanderthals.*

German Anthropologist Weinert, among others, has noted that extreme Neanderthals had a "large and heavy head, exceeding in dimensions the limits of variation found in modern man." That is clearly an evolutionary trait that distinguishes Neanderthal from Sapiens, and it would be normal to expect that it would have progressively increased as the two forms diverged.

Alika Lindbergh's reconstruction of the living pongoid man. Only the hair, the shape of the ears and the eyeballs were inspired by eyewitness descriptions of live Asiatic wild men.

Let's now focus on the face.

Let's look at a picture of the specimen's face from Broca's plane,[54] or a representation of its facial appearance from the same perspective. Let's now superimpose, in dotted lines, the outline of the skull of the Chapelle aux Saints Man seen from the same direction, and scale it up so that it has the same facial height (nasion–gnathion, i.e. nose to crown). The coincidence of the two faces is clear in the smallest details. The width of the orbits, the bi-zygomatic width of the face, the position of the sub-nasal point and of the mouth, the contours of the mandible, and the top part of the head.

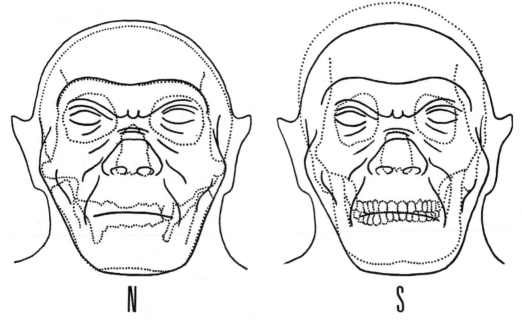

The essential features of the specimen's face, superimposed (dotted lines) on the Neanderthal skull of the Chapelle aux Saints (N), and on the skull of Homo sapiens *(S).*

If we follow the same procedure, but now use a composite sketch of *Homo sapiens* skulls fitted to the same facial height, we cannot find the same degree of coincidence, no matter which points we compare. The upper skull is always too high, the eyes are not far enough apart, the cheekbones are not prominent enough, and the jaw is weak.

What is most original in our specimen's face is undoubtedly his nose. But is it a Neanderthal nose?

Certainly not if one is to judge on the basis of the numerous reconstructions proposed for the appearance of the typical paleanthrope, which is usually shown with a wide nose with large nostrils, like a black African, or a curved nose like a Papuan man; he is sometimes also pictured with a Bourbon nose, or a Jewish nose. It's a wonder he wasn't imagined with Cleopatra's or Cyrano's nose. But no one ever

---

54   With such an orientation, the ear canals are at the same level as the sub-nasal point.

thought of giving him a snub nose!

Ah, but yes! There is indeed one such representation by Dr. Faure, a physician from Nice, which appeared in 1923 in a relatively obscure publication, since neglected. Its author claimed that it should be considered "not only as the most complete, but also the most exact from an anatomical and physiological point of view, within the degree of relative certainty appropriate to the biological sciences." Furthermore, there is also a sculpture by a German artist, R.N Wegner, where the Neanderthal sports an upturned nose. However, in the May 1968 issue of *Natural History*, where it appears as an illustration for an article by professor C. Loring Brace, the legend contrasts it with reconstructions deemed more acceptable and labels it a caricature.

So, what was it really like?

Thanks to the Chapelle aux Saints, the Gibraltar, the La Quina, and the Krapina skulls, among others, in which the nasal region is well preserved, the anatomy of the Neanderthal nose is well known, and is immediately remarkable by its originality.

In their classic text *Les Hommes Fossiles* (*Fossil Men*), Boule and Valois wrote that: "...the nose of *H. neanderthalensis* is not at all like that of the anthropoid apes and is more different from it than is the nose of modern man. While this fossil man is closest to the apes in many characteristics, it is most different from them in its nasal region, which rather than being ape-like is more *ultra-human*." Professor Etienne Patte quite appropriately described one of these *ultra-human* traits: "The nasal spine is highly developed ... it points both forwards and upwards." Furthermore, Robert Broom, the prominent South-African anthropologist, remarks first that "... in no Australian does one find eyebrows comparable to that (Neanderthal) type, nor such a jutting nose," and he continues: "The ample development of the nasal region is not a sign of a primitive type, but most likely a late specialization."

So how does that exceptional nasal appendix present itself?

Guerasimov spent years developing an acceptable method for reconstructing the nasal profile solely based on an examination of the skull, more specifically to fix the exact position of the tip of the nose. "The profile of the nose is determined by two straight lines the first tangent to the last third of the nasal bones, the other a prolongation of the nasal spine. The intersection of these two lines will generally determine the position of the tip of the nose."

I have used a similar method that gives practically the same results. I borrow from Boule and Valois a well-known figure that compares the skulls of a chimpanzee, an average Frenchman, and the Chapelle aux Saints Man. On each one of those let's imagine on the one hand the prolongation of the nasal bones and on the other the curve suggested by the nasal spine; one then finds a rough profile of the creatures to whom belonged those skulls.

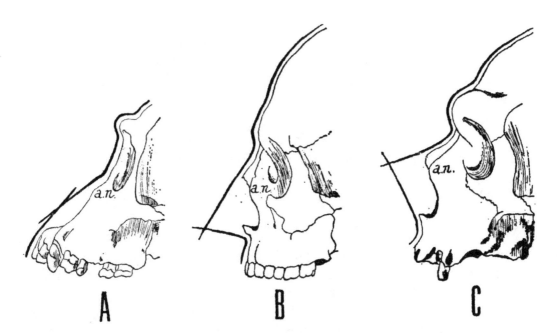

Reconstruction of the nasal profile of A: a chimpanzee; B: a modern Frenchman;
C: the Chapelle aux Saints Neanderthal. (From Boule and Valois, 1962)

In extreme Neanderthals, the deepening of the root of the nose, the development of the nasal apophysis of the maxillary bone, is so pronounced that it tends to lift up the nasal bones until they are nearly horizontal, and finally the upturned position of the nasal spine probably endowed the nose with a strangely upturned appearance, without equivalent in today's races of *Homo sapiens*. The nostrils most likely opened directly forwards. These are exactly the features we observe in our specimen.

One might wonder why the various authors who have tried, often with great care, to reconstruct the fleshy parts of Neanderthals, practically never managed to reconstruct the nose suggested by the bony structure of the skull. The upturned character of the nose is barely suggested in the reconstructions by Schaafhausen, Mollison and Friese, Mascré and C.S. Coon, and is completely ignored in those of Boule and Joanny-Durand, Roubal, Heberer, Wandel, Knight, McGregor, Jaques, Brink, M.P. Coon, Sinelnikov and Nestourkh, Wilson, and Augusta and Burian. Even Guerasimov did not apply his own method correctly. Why? It's only Faure and Wegner, as I mentioned, who have shown an exact representation of the upturned nose of the extreme Neanderthal. However, Faure's reconstruction remained unknown and Wegner's was taken as a caricature.

One is led to suspect that it is only the fear of ridicule, rather than incompetence in anatomical matters, that prevented most anthropologists from attributing to the classic Neanderthal a nose so grotesque that it would irresistibly provoke laughter.

One last point about that nose. Since the ice age paleanthropes were adapted to extremely low temperatures, one might be surprised to find that our specimen, as a result of their continuing evolution, would have such a large nose and large, round, wide-open nostrils. Actually, in contemporary human races, nostrils tend to be narrower in cold climes and wider in warmer regions, and also smallest when the air is dry and widest when it is humid.

These are well-established facts, in agreement with physiology. But we know that the Neanderthals had, via the greater development of their sinuses, resolved in a completely different fashion the problem of warming up and humidifying the air that they inhaled. The rules that apply to *Homo sapiens* are thus no longer valid in that matter.

Who knows whether the peculiar form of the nose and the nostrils might not be, among specialized Neanderthals, an adaptation to the cold? Of all primates, that which it most resembles in this respect is without a doubt *Rhinopithecus*, which is one of the monkeys best adapted to low temperatures. It lives in a mountainous area of China and Tibet, often snow-covered for more than half the year, and it is often found at such high altitudes that the Chinese call it "the snow monkey."

The upturned nose of the former Neanderthals, as well as that of our specimen, was surely related to an acute development of the sense of smell. Australian aborigines, some of whom exhibit that kind of nose,[55] use their sense of smell a lot. They are with their nose in the air, sniffing to find water holes, their prey, or potential enemies. It's obvious that to have such a strongly olfactory perception of one's environment, large nostrils opening forward are particularly appropriate.

After the nose, let's consider the mouth.

In a number of reconstructions, the classic Neanderthal is represented with the thick, turned-up lips of black Africans. Some thoughtful authors, Professor F.S. Hulse among others, have noted that "this would seem quite inappropriate for an Arctic creature"; Professor J.H. McGregor remarked that "incisors set vertically end-to-end when closing make such lips rather unlikely."

It's much more likely, given the climatic conditions as well as their dental structure, that Neanderthals of the last glaciation had no lips at all and a widely stretched mouth. Again, that's exactly what we find in our specimen.

Professor Howells has also noted that in those Neanderthal skulls "the teeth were robust and slightly larger than ours." And isn't that what we also notice in those few photos that show our specimen with his mouth open?

55   See for example the picture of a Tiwi from Melville Island, published by Coon (1962, facing page 84), or the photos of Australians from Cape York Peninsula and Murray River, published by Hooton (1946, plate26), as well as the photo of two Tasmanian women published by Nestourkh (les Races Humaines, p.90).

And now, let's consider the limbs.

The upper limbs of extreme Neanderthals are characterized by three salient traits.

The first one is the relative shortness of the forearm with respect to the upper arm, a trait we also observe in our specimen.

The second salient trait is the degree of muscular development of the forearm, which greatly exceeds that of the upper arm, a phenomenon linked in part to the pronounced curvature of the radius and the decreased curvature of the ulna.[56] The huskier Neanderthals had arms like those of Max Fleisher's hero Popeye, the brawny comic strip sailor.

This specifically Neanderthal development of the forearm probably explains the unusual curve of the specimen's left arm, which I originally attributed to a fracture of the ulna, there being an open wound on the outside of the forearm. Perhaps that fracture, if it really was total, only accentuated an anatomical structure unfamiliar to anyone used to the anatomy of Modern Man but which would be perfectly normal in a Neanderthal.

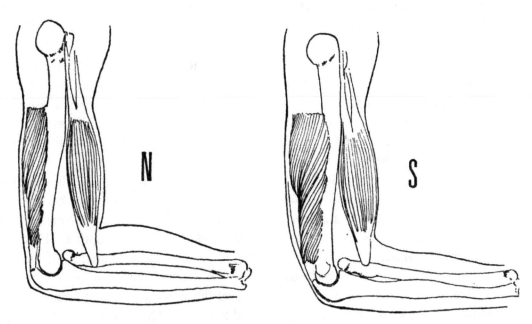

Reconstruction of the arm and forearm muscles of a Neanderthal (N) compared to that of
Homo sapiens *(S). (After Loth, 1938)*

---

56   In one of his studies on the reconstruction of the fleshy parts of Neanderthals, Loth wrote, in 1936: "The differences in the length of the olecranon apophysis and the position of the tuberosity of the radius lead us to suppose that the arm's extensor muscles (triceps) and its flexers (biceps) were smaller in volume among Neanderthals because of the different length of leverage, but nevertheless just as strong. In the forearm on the contrary, the curvature of the ulna and of the radius and their spreading in the presence of a larger inter-bone tendon must have resulted in larger insertions and much larger muscles."

The third noticeable trait of the upper limbs in Neanderthals is the relative size and width of the hand, which is particularly striking in our specimen.

The hand, a most important organ in the evolution of hominids, deserves special attention.

That belonging to the classic Neanderthal presents characteristics whose full importance was not always grasped, due to a lack of sufficiently numerous, and especially entire anatomical remains, which were sometimes wrongly interpreted.

Because the wrist and palm bones (carpals and metacarpals) have usually been best preserved, all authors agree on the relative size and width and on the compact and thickset nature of the Neanderthal hand. Disagreement has focused on the structure and the function of the fingers, in particular the thumb.

Many anthropologists have long emphasized the lesser opposability of the thumb among typical Neanderthals. The ability to oppose the thumb to the other fingers essentially depends on the articulation at its base. It is thus linked to the shape of the lower extremity of the metacarpals, articulated on a specific carpal bone, the trapezium. In humans, that contact junction is in the shape of saddle, which greatly improves the ease of torsion of the thumb.

In Neanderthals, that surface is often convex or even flat, which makes the articulation stiffer; the thumb is less readily opposable but can move further sideways.

This is clearly a primitive trait. In human embryos, the thumb does not yet have a saddle-shaped joint, and until they are six months old, babies do not use their thumbs to grasp objects. This arrangement is at the extreme opposite of that found in anthropoid apes, whose hand is built to grasp branches as firmly as possible, with a clearly opposable thumb.

Some have rebelled against the theory that the hands of Neanderthals were less prehensile than ours. This idea bothers those who imagine that Neanderthals are our ancestors and we ultimately descend from apes: if that were the case, Neanderthals should be closer to apes than we are. It was pointed out that in both La Ferrassie (Dordogne) specimens, the male as well as the female, the articulation of the first metacarpal on the trapezium is saddle shaped. Dr. Coon went as far as to conclude that: "Overall there is no proof that the hands of the western Neanderthals were any different from those of today's European hand laborers." The delicate nature of some Mousterian stone tools seems to favor that view.

However, we finally have to admit what common sense should have led us to suspect a long time ago: the Neanderthals that dispersed across the globe for tens of thousands of years did not all reach the same level of evolution. In the same way, today, some representatives of *Homo sapiens* are still living in the Stone Age, while others find a way of walking on the Moon.

If there lived at La Ferrassie some Neanderthals with more agile hands than the Chapelle aux Saints Man, capable of very delicate work, there were also some with much clumsier hands, for example, near Simferopol, in Crimea. In his detailed 1941 study of the "The Hand of the Fossil Man of the Kiik Koba Cave," Soviet anthropologist G.A. Bontch-Osmolovski showed that that hand was more primitive than that of most Neanderthals of Western Europe (and also closer to the embryonic form common to all hominids). This is, of course, in prefect agreement with the coarseness of the flint tools associated with the remains of classical Neanderthals. All these characteristics distance them even further from *Homo sapiens*.

A common phenomenon in evolution, already noted by Goethe in his famous Compensation Principle, is that it is among the most specialized types that the hand remained the most primitive.

In any case, the weak opposability of the thumb among extreme Neanderthals sheds some light on the minor amount of twisting of the thumb seen in our specimen. It also accounts for its thinness since a non-opposable thumb need not be flattened.

The unusual length of the specimen's thumb is also striking, reaching almost all the way to the first articulation of the index finger. Most authors, on the other hand, agree in thinking that Neanderthals had rather short thumbs.

But this opinion is not unanimous. Writing about the same specimen, the Taboun female-I, Dr. Coon said: "…her thumb is short," while Professor Piveteau, on the contrary, spoke of "large and wide hands, with a particularly developed thumb."

To settle the issue, I carried out a comparative study of some of the most complete Neanderthal hands available, namely those of the females at Taboun, and La Ferrassie, and of the males of Kiik-Koba and Skhoul V.

What is most striking, when one compares these hands with those of *Homo sapiens*, is the strong curvature of the thumb's metacarpal bone, resulting in its somewhat spreading away from the other fingers and widening the palm of the hand.

A careful reconstruction of each one of these incomplete Neanderthal hands in the light of the others leads to the creation of a synthetic Neanderthal hand (C), and reveals a characteristic trait: the greater relative length of the thumb. Measuring it along its curvature, it turns out to be almost as long as the combination of metacarpal II and phalange II next to it. The only hand I looked at which is not like that is that of the Skhoul specimen. But that should not be surprising for a specimen that is so unrepresentative of the Neanderthal lineage that it is suspected to be a *Homo sapiens* hybrid. The study of his thumb suggests that he might actually be the result of a Sapiens mother with a Neanderthal father.

Bontch-Osmolovski has wisely pointed out the similarity between the hand of the Kiik-Koba Man and that of a human fetus. In fact, one should have logically expected to see a rather long thumb in the primitive Neanderthal hand, since for all primate groups, the thumb is always relatively longer in the embryo than in the adult.

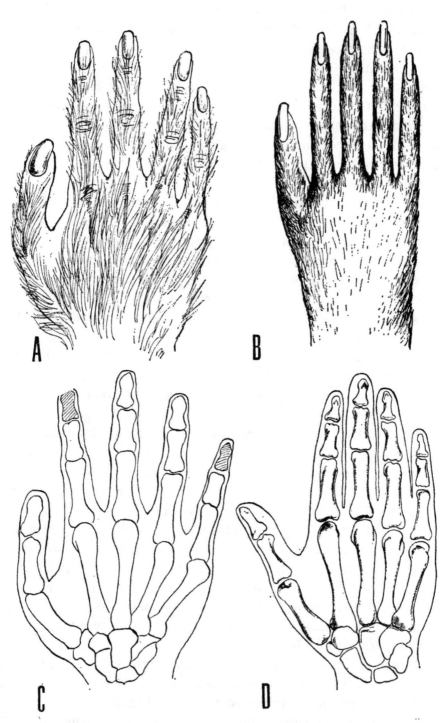

Comparison of the hands of A: the specimen; B: the hand of the Ksy-gyik (after Khakhlov);
C: synthetic Neanderthal; D: Homo sapiens.

This feature is clearest among Neanderthals; Modern Man comes next. It's among monkeys that the adult thumb is most reduced, especially among the specialized tree-dwellers, whose hand has a tendency to narrow and transform into a hook.

Their hands suggest the following evolutionary sequence among hominids: Neanderthal – Sapiens – Gorilla – Chimpanzee – Orangutan.

One must understand that this series simply indicates the general direction of the evolution of the hand and certainly does not imply any filiations. It's not even a sign of progress. It simply appears as if at every step of the hand's gradual transformation, its owner took to best advantage the structure at its disposal and perfected it by adapting to the specific needs of its own species. The Neanderthal best used his large hand with its long weakly opposable thumb to grasp the rock ledges it had to climb and to gather large amounts of plant food. *Homo sapiens*, with his mobile thumb, still long enough to oppose itself exactly to the tip of the other fingers, had exactly what he needed for precision work and to become the artisan of the Primate world. Gorillas and chimpanzee still have opposable thumbs, but they are now too short to assure a delicate fingertip grip; their hand has become an ideal clamp to hold on to branches. (Monkeys have such a need for that kind of grip to ensure their safety in trees, that even their feet have been transformed in that direction.) In Orangutans, the hand, with its atrophied thumb and long curved fingers, has become a large hook, ideal for arboreal acrobatics.

Coming back to our specimen, we conclude that its hand, in all its characteristics (width, reduced opposability, relatively longer thumb, which is also thin and laterally displaced from the other fingers) fits perfectly in all its smaller details the hand of a specialized Neanderthal.

We shall see that this is also the case for our specimen's feet.

The feet of classic Neanderthals are well known, and not only from skeletal remains, but also from footprints they left in the La Basura cave, Toirano, Italy. The study of bones and footprints reveals that their feet were rather short and wide, in the sole as well as the toes; overall then, squarer, flatter, and more massive.

Of all the known Neanderthal feet, the most characteristic is obviously, as one would expect, that of the Kiik-Koba Man, a more extreme Neanderthal than the extreme ones of Western Europe. This foot has been studied in great detail by Bontch-Osmolovski and Bounak (1954).

One of the most striking features of the restored Kiik-Koba foot is the way its toes fan out, jutting toes II and III forward. This is in sharp contrast with the placement of the toes in *Homo sapiens* where toes go from long to short from I to V (even if toe II is sometimes longer than the big toe).

Dr. Coon notes that "… this foot could well have left the prints found by Baron Blanc in an Italian cave." However, when one superimposes the skeleton of the foot of Kiik-Koba in one of the footprints of the Toirano cave, one finds that the metatarsals should be even more spread apart than they are on the skeleton mounted by Bounak.

Pales (1960) studied the Toirano footprints and deduced some characteristics of the feet that made them, and even of their gait: "The tarsal area is relatively short. The first metatarsal bone is flatter and less twisted (by 10°) than that of a modern European, with the result that the big toe is more spread out... Furthermore, the anatomical axis goes through the second or third toe, or between them.... The fifth toe is slightly twisted about its axis." Pales also noted that in one of the tracks, the big toe repeatedly left a round rather than oval depression, and he thinks that, unless that might be due to some malformation or accident, that toe had "stepped in a clay that was more malleable than elsewhere around it."

Actually, that flexure of a toe might reflect a physiological tendency normal in all toes. The highly curved nature of Neanderthal toes has long been recognized. According to the Darwinian idea of evolution, this feature was attributed to their greater ability to climb trees, being closer to our simian ancestors.

Albert Gaudry, a prominent anatomist, demonstrated a long time ago that man's plantigrade foot is a primitive structure and can in no way have developed from a specialized organ such as the monkey's foot and transformed into a grasping hand.

After analyzing the 63 characteristic trait of the Kiik-Koba foot, Bontch-Osmolovski found that 26 of them were no different than those of a modern human's foot, 25 are closer to those of anthropoids, but there are also 12 that are even further from the feet of apes than from those of humans, in other words, "ultra-human" traits. The evolution of that foot actually diverges from that of tree-dwelling apes, as well as from that of Modern Man, a savanna runner, because it has a completely different function. As Bounak emphasized, it is the foot of a rock climber. That foot has to be as prehensile as possible to get a good grip on rocky terrain but in a different way than for grasping branches. Those traits that resemble those of the anthropoid foot are simply the result of a convergent evolution.

In any case, our specimen's foot turns out to conform completely to that of a specialized Neanderthal: relatively short, quite wide, short curled toes fanning out so as to put the second and third toes forward, slightly spread out big toes, and inwardly twisted little toes.

One point to note regarding the width of our specimen's foot. Based on the value of the crucial (width/length) index, it is even more impressive than in the Kiik-Koba Man. The specimen's foot is even further in this respect from the Kiik-Koba Man than the latter is from the average Frenchman's foot. But isn't that to be expected in a creature that has continued to evolve in a direction taken by its ancestors?

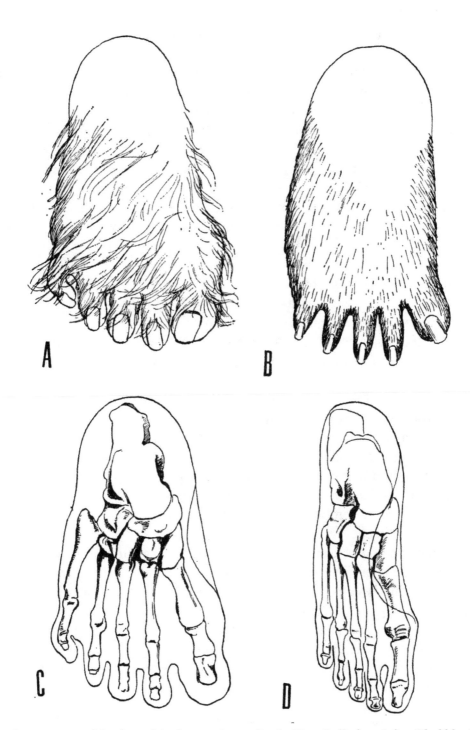

Comparison of the feet of A: the specimen; B: the Ksy-Gyik *foot (after Khakhlov);
C: the skeleton of the Kiik-Koba Neanderthalian, inserted within the contour of a Toirano
footprint; D:* Homo sapiens.

To sum up this comparison between the anatomy of our specimen with that of Neanderthals, we have to conclude that we find in it the whole mosaic of specific Neanderthal characteristics, each one of which having reached a more or less elevated level with respect to the mosaic of *Homo sapiens* traits. At best, we note that certain traits of our hairy man, for example the absolute increase in height, the relative increase in the size and the height of the head, the width of the hands, and especially the feet, have continued their evolution beyond the stage reached in the early Würmian age.

Is there any reason that we should not identify the specimen as a late representative of paleanthropian people? On the contrary. Only if we had not made these remarks would we might have had some grounds to be surprised, concerned, or astounded about the authenticity of the frozen man. For the evolution of various characteristics never proceeds at a uniform rate within a group: any specialization is linked to rapid changes in some structures while others change more slowly or not at all. That is what is called evolutionary allometry. It would have been abnormal for surviving Neanderthals not to grow taller, as it would also have been abnormal for them to merely transform into exact enlarged replicas of what they were.

Such are the obstacles that would have stood in the way of a hoaxer who would have tried to construct a false Neanderthal, soon to be found out by the more clever specialists.

What hoaxer could possibly have thought of creating such subtle anatomical details as those revealed by a careful study of its anatomy? There is one, the upturned nose, which although suspected by some anthropologists, never gained recognition, in spite of arising as a mechanical consequence of the underlying skeletal structure. Others, such as the thin, spread-out and weakly opposable thumb, or the crooked toes, adapted to climbing mountains, are poorly known and even controversial. And finally, the hairiness is rarely acknowledged, although it is a logical adaptation to living in a harsh climate.

Which anthropologist, as clever as erudite and exceptionally well informed, could have designed the plans for such a model? Why would such a remarkable man, who would undoubtedly have been a star in his field, have lowered himself to perform such a hoax?

There is no more striking proof of the authenticity of the specimen than its perfect conformity not only with the anatomy of the more specialized Neanderthals but with the anatomy they would have acquired following the subtle laws of biology, if they had been able to pursue their evolution until today.

Reconstruction by Alika Lindbergh of the pongoid man as a living creature.

Chapter 11

A HISTORY OF MAN-BEASTS
From pre-Neanderthals to Pongoids

"Dripping saliva, stinking and hairy…
endowed with tremendous strength, he haunts
the fens and the forests, where neither beast
nor men dare to venture."
— Beowulf, Portrait of Grindel, the wild man,
in "Beowulf," 4th Century Anglo-Saxon poem

That the hairy frozen man exhibited for three years in the United States is truly a Neanderthal in the broadest sense is no longer in doubt after the careful analysis of its external anatomy.

Unless he's the only one of his type.

Although this is as unlikely as seeing a monkey randomly typing a page of Shakespeare on a typewriter, it is *theoretically* possible. The specimen could be a man of our species who, through highly improbable genetic combinations, would have inherited from the remote ancestor of Sapiens and Neanderthals all the traits characteristic of the latter, and only of the latter. Or, even more fantastic a hypothesis, he might be the result of the forbidden love between a woman and an anthropoid ape, or vice-versa. The combination of the two chromosomal sets, however dissimilar, might have produced such confusion in the genes that, by pure chance, *all* the Neanderthal traits would have appeared through a mutation.

Don't laugh! Such explanations have been advanced seriously; anything rather than admitting the possible survival of Neanderthal Man today.

Just to calm down the most punctilious scientific critics, it remains to be shown that our specimen is not an isolated case but that he is a member of a population of similar creatures. For those who have not had access to the work of my Russian colleague and colleague Boris Porshnev, I will present a control test.

Among the vast harvest of information gathered by his Russian team, let us select a few descriptions sufficiently detailed to allow a comparison with that of our specimen, and also from areas far from each other. It immediately appears, not surprisingly, that precise relationships result from the comparison of captive specimen and corpses. As for the anatomy of the foot, comparisons may be based on the study of very clear footprints. Furthermore, equally useful descriptions arise from the synthesis of a number of partial descriptions that complement each other.

First things first. I first selected the description of the Ksy-Gyik of Dzungaria, presented by zoologist Vitali Andreyevitch Khakhlov and based on the careful analysis of two captured specimens. This selection is particularly appropriate because it is illustrated but I must add a few words.

Khakhlov's drawings are not to be taken as realistic representations. They are reconstructions based on the oral descriptions of eyewitnesses, without the accuracy of modern police facial reconstructions based on photographic techniques. As in a caricature, the most salient traits have been exaggerated to make them more noticeable. What is most striking in a creature that looks like Man, is what distinguishes it from the latter, namely the more animal and more specifically simian characteristics. Making allowances for the more outrageous exaggerations, Khakhlov's sketches look a lot like caricatures of our own specimen and emphasize its original anatomical features. The text faithfully quotes Khakhlov's informers and makes it possible to understand the exaggerations and in some cases to correct Khakhlov's interpretations.

The second description I selected is that of the Mongolian Almas, as it was summarized on the basis of numerous reports by my friend, the Mongol Academician Rintchen, of the Yöngsiyebü clan.

The third description worth drawing on is that of the Goul-biabane of the Pamir, presented by Mikhail Stepanovitch Topilsky, a retired general who in 1925, as Porshnev relates, was able to conduct a careful study of a corpse in the eastern Pamir area.

The fourth and last description is that of the Kaptar, from the Caucasus, which lies halfway between a detailed individual description and a synthetic portrait. It is mainly based on the examination of a captive specimen by Vazguen Sarkisovitch Karapetian, a physician and lieutenant colonel, again as related by Porshnev. For the details of the foot, the description is enriched by the information collected by V.K. Leontiev, wildlife inspector with the Hunting Department of the Autonomous Soviet Socialist Republic of Daghestan who, as an expert in the local fauna, studied in detail the footprints of a creature he encountered. Finally, other anatomical details are provided by the collective representation compiled from more than three hundred first-hand reports by Dr. Marie Jeanne Koffmann.

I could not possibly reproduce verbatim the documents on which these four descriptions are based; they would overwhelm this book. Khakhlov's text alone is more than 25 pages long. Conscientious skeptics will have to consult the original documents. The essence of these observations is presented in the Summary Table of Descriptive Traits that follows. The most characteristic features of the anatomy of the classic Neanderthal, of our frozen specimen, and of the four wild men from different parts of Asia are all compared. For comparison, the existence of such features in Modern Man is also included.

...table key and text continues on page 196

DESCRIPTIVE TRAITS	EXTREME NEANDERTHALIAN	FROZEN SPECIMEN	KSY-GYIK	ALLMASS	GOUL-BIABANE	KAPTAR	HOMO SAPIENS
Very hairy all over	?+	+	+	+	+	+	-
Skin visible through hair (as for anthropoid apes)		+		+	+	+	-
Naked face (or beard or mustache)		+	+		+	+	+/-
Sparce eyebrows		+	+				-
Small hairs scattered on face		+			+	+	-
Longer hair on top of head			?+	+	+	+	+
Less hairy on knees		+	+		+		-
Dark colored skin		-	+		+/-	+	+/-
Massive head, very large face	+	+	+			+	-
Long head, back to front	+	+	+				+/-
Strongly sloping forehead	+	+	+	+	+	+	-
Prominent eyebrows	+	+	+	+	+	+	-
Eyes very far apart	+	+					+/-
Dark or brown eyes			+		+	+	+/-
Sticking out cheek bones	+	+	+		+	+	-
Pointed ears		+	+		+		-

DESCRIPTIVE TRAITS	EXTREME NEANDERTHALIAN	FROZEN SPECIMEN	KSY-GYIK	ALLMASS	GOUL-BIABANE	KAPTAR	HOMO SAPIENS
Long ear lobes			-		+		+/-
Extremely large nose	+	+				+	-
Very upturned nose	+	+	+		+	+	-
Large nostrils opening forward		+	+				-
Lack of sino-labial groove		+					-
Very wide mouth	?+	+	+				-
Absence of lips	?+	+	+				-
Extremely large and strong teeth	+	+	+		+		-
Prognathous jaw, flat muzzle	+	+	+	+	+		-
Massive, narrow, rounded mandible	+	+	+		+	+	-
Chin small or absent	+	+	+			+	-
Forward leaning stance	+		+	+		+	-
Head sunken in shoulders	+	?+	+				-
Heavy and powerful nape	+	?+	+				-
Very arched back	+		+	+		+	-
Very wide shoulders	+	+				+	-
Deep, nearly cylindrical chest	+	+				+	-

DESCRIPTIVE TRAITS	EXTREME NEANDERTHALIAN	FROZEN SPECIMEN	KSY-GYIK	ALLMASS	GOUL-BIABANE	KAPTAR	HOMO SAPIENS
Very long trunk	+	+	+				+/-
Long upper limbs	+	+	+		-	?+	+/-
Short forearm/arm	+	+					+/-
Extremely large hand	+	+	+				-
Rather wide hand	+	+	-		+		+/-
Very long fingers	+	+	+			+	+/-
Thumb long, narrow, spread out	+	+	+				-
Weakly opposable thumb	+	+	+				-
Fingernails narrow and domed		+	+				+/ -
Short lower limbs	+	+	+			?+	+/-
Legs twisted or bent knees	+	?+	+	+		+	-
Legs short/thigh	+	+					+/ -
Short foot	+	+	+			?+	+/-
Fanned toes (axis between toes II and III)	+	+	+			+	-
Crooked toes	+	+		+		+	-
Spread out big toe	+	+	+	+		+	+/ -
Toes about all equal size	+	+	+	+		+	-
Small toe curved inwards	+	+		?+			-
Toe nails narrow and domed		+	+				+/-

In this table, the + sign indicates the presence of the character under consideration (be it a positive feature, like a "sub-orbital swelling," or a negative one such as "lips missing"); the - sign , the absence of this character. When a question mark precedes the plus sign (? +) it means that the presence of the character is most likely. The sign (+/-) means that the character under consideration is sometimes present, sometimes not, or if one prefers, that it is sometimes found (in some individuals, or in some races). The absence of any sign reflects the lack of information on that particular characteristic.

Of course, those characteristics common to all, such as human height, bipedal stance, or pectoral breasts, do not appear in that table; only those characteristics which might allow identification or differentiation.

An analysis of the Summary Table reveals, on the one hand, the striking correspondence of descriptive characteristics between the extreme fossil Neanderthals, our frozen specimen, and the various hairy wild men of Asia (whether belonging to an isolated population in the Caucasus or disseminated within the vast mountainous area of Central Asia) and, on the other, the number of common traits among those creatures, which are found only in some individuals or certain races of our species, and which are most often completely absent from it.

Of the 53 characteristics considered, 37 (more than two third) clearly distinguish *Homo sapiens* from all the others and only 19 are found in those others that may sometimes exist in our species. Only one trait, that of abundant hair on the head, is common to all, besides of course the numerous anatomical characteristics common to all hominids, which are not included in the table.

That being said, one nevertheless notes some contradictory features in the table. But their importance is minor.

First of all, there is the unusual pallor of our specimen's skin. As I mentioned earlier, it might be attributed to a long period of captivity in the dark. Or perhaps is it a consequence of its posthumous immersion in the freezer. The Kalahari Bushmen are usually described as very dark, although their skin actually resembles the hue of tanned leather. What happens is that in a region where water is a rare and valuable resource, they practically never wash. The discordance between the skin color of our frozen man and that of the other troglodytes might be due to the combined effect of two factors: the pallor of a prisoner in the former, an excess of dirt in the latter.

The absence of earlobes in the Ksy-gyik contrasts with the elongated earlobes of the Goul-biabane. This detail, which in our species is the subject of considerable racial and individual variation, is of no value as a criterion for differentiation.

Finally, the upper limbs are described as long to very long in all non-Sapiens, except for the Goul-biabane and perhaps also the Kaptar. This difference might be explained by the confusion that occurs in most languages between those words that denote the whole limb, the limb without its extremity (hand or foot), the upper part of the limb, or even its extremity. In French, the word *bras* might refer to the whole upper limb, the limb without the hand, or only the upper arm. The more trivial term *patte* may refer to the whole limb or to only to its extremity. In Russian, *rouka* denotes the hand as well as the whole arm. Finally, the relative proportions of the upper limbs are particularly equivocal among Neanderthals; it's quite long when taken as a whole, but relatively short if its extremity is excluded, because its greater length is due entirely to the enormous size of the hand.

Finally, one must recognize that the Summary Table shows a nearly absolute agreement of descriptive characteristics in all types included, except *Homo sapiens*. One can then confidently fill in the gaps that appear in the table because of the lack of information.

One may then conclude that, overall, fossil Neanderthals of the extreme type must have looked pretty much like our specimen or like the existing "wild men" of Asia, and that, more specifically, our specimen must also have had long hair, dark eyes, and, when alive, a forward leaning stance with bent knees. One should also believe that the Ksy-gyik's skin should be visible through its hair, have eyes set wide-apart, crooked toes, and lack of a naso-labial groove; furthermore, that the Almas of Mongolia is most likely to be very similar to its better-described immediate neighbor in Dzungaria; and that this is also the case of the more southerly Goul-biabane, even of the Kaptar nearer Europe.

One should note that there is a family of anatomical traits that does not appear in the Summary table: that which relates to the hair, namely its color, length, and texture. Its omission is justified by the great variability of these characteristics among any species, as well as the often-subjective nature of their description in common language. Thus, for the shepherd of a yak herd, 5 cm [2 in] long hair are described as "short," while the same hair will appear "long" to a cowboy used to horses.

For even more obvious reasons the Summary Table does not include data on height. Stature is usually compared to that of the inhabitants of the region, but in some cases is said to be "half a man's height" while in other cases it is said to be 2.10 to 2.20 m [approx.7 ft]. One can't attribute too much importance to such reports for now. On the one hand, any one individual goes through a range of heights over a lifetime. On the other, an encounter with a wild man, built like Hercules and looking larger because of its hairiness, makes a very strong impression. To quote a Russian proverb: "Fear has big eyes."

It is quite possible that sometime in the future the study of a sufficiently large sample of Asiatic wild men, photographed or captured, will allow distinguishing local races, differing in their stature, hair color, or texture. In that case, the existence of such races will most likely be linked to the isolation of small groups of individuals in certain mountainous geographical pockets, such as the Caucasus, the Tsinling-Chan and Houn-He mountains of China, and the Annam chain in Vietnam. Where such isolation is most ancient, one would expect a pronounced amplification of characteristic Neanderthal traits, such as the size of the head and its back-front lengthening, the development of the eyebrow ridge and the occipital bun, the width of the foot, etc. It might even be possible to discover some signs of degeneration; that is how Dr. Koffmann accounts for the presence of albino Kaptars in Azerbaijan. It would also be natural to expect racial differences to have developed in response to the particular ecological environment where these relic hominids have lived. Organisms evolve differently in tropical jungles than in frozen tundras, or in mountains than in plains.

Searching for geographical varieties of pongoid Neanderthals is rather premature at this point. My aim here has been only to establish that: (1) the specimen I have studied is similar in all its visible anatomical characteristics to the extreme type of fossil Neanderthals; (2) that specimen is not unique, but part of a population of similar beings dispersed today through a large area of the paleo-arctic region and spreading into the oriental region of Southeast Asia; (3) that today, hominids similar by their anatomical characteristics to the classic late Würmian Neanderthals, but with more pronounced characteristics, have survived within the former range of distribution of Neanderthals, now broken up and much reduced in area.

The survival of paleanthropes to this day may seem contrary to available paleontological data; the most recent remains of Neanderthals are supposed to date from the late Würmian.

However, that would be to forget that a fundamental truth of paleontology is that "a negative fact means nothing."* And besides, positive paleontological data, some indisputable, have long shown that Neanderthals not only survived beyond the late Würmian (upper Paleolithic), but, after the melting of the ice caps, into the Neolithic and even in the Bronze Age.

Already in 1908, Kasimiersz Stolyhwo, the prominent Polish anthropologist, described a whole series of Neanderthal skulls dating from historical times. Thus, near the Novosiolka (near Kiev) skull, were found chain mail and iron weapons; according to Stolyhwo, this skull would have belonged to a man living in recent times. As to the Poszuswie skull, he dated it at the 10th century of our era. These studies were immediately the butt of violent attacks and, one must admit, were thoroughly discredited.

At the time, it's true, knowledge of Neanderthals was still fragmentary, and most anthropologists stuck fanatically to the idea that man descended directly from the apes. As Neanderthal Man presents some ape-like traits, he was unanimously considered as the direct forebear of modern Man and *must logically have come before it*. Hence, each time clearly Neanderthal remains were found, they were automatically attributed to geological strata more ancient than those where the first human remains

---

*   Or: The absence of evidence is not evidence of absence! — The Editor

had been discovered; and, conversely, when remains of Sapiens were dug up, they were attributed to more recent deposits. As the means of dating were then far from accurate, it often happened that Neanderthal remains were incorrectly made older and Sapiens remains made younger.

Once the dating techniques were improved, other stratagems were invented to remain within the Darwinian framework, although the concept of evolution as a spreading bush instead of a linear sequence had already been generally adopted. Whenever the reliably dated deposit where Sapiens remains were found was thought to be too old, it was assumed that they must have been buried there at a later time. And whenever Neanderthal remains were found in strata thought too young, it was assumed that they did not belong to a *real* Neanderthal, but to a *Homo sapiens* "having conserved Neanderthal or Neanderthaloid traits." Another argument used by anthropologists faced with remains that, according to some preconceived idea, do not fit with the age of the deposits, is to claim that "they cannot be dated."

A significant example of how facts can be systematically distorted to fit them into some dogmatic conception of human origins is the story of the Neanderthal skullcap discovered in 1918 near the Podkumok River, in the northern Caucasus. Given the highly developed brow ridge and its sloping forehead, there could be no doubt as to the identity of the creature to which it belonged. The slope of its forehead is at least as strong as that found in generalized Neanderthals. The skull's profile was practically identical to that of the pre-Neanderthal woman of Taboun. Seen from above, the contour of the fragment of skullcap is practically identical to that which characterizes any classic Neanderthal. The nature of the owner of that skullcap was not even disputed, as it was believed to originate from the early Würmian, as it had immediately been suspected. However, that was very far from its actual age.

In a 1966 article entitled "The Origin of Man and of the Hairy Hominoids," Porshnev quite appropriately commented on that same erroneous dating which, for him, brought *concrete proof of the survival of Neanderthals*, at least to the dawn of historical times: "In the History of Science, one scientist's mistake may sometimes merit a monument. Such was the error made by the Russian geologist V.P. Rengarten (1922). Without bothering to visit the site, he attributed the geological deposit where the 'Podkumuk skullcap' was found at Piatigorsk in 1918 to the Würm glaciation. That geological date provided anthropologists such as Gremiatsky (1922), Zaller (1925), Weinert (1932), Eickstedt (1934), and others the assurance that the Podkumuk skull indeed presented an ensemble of Neanderthal morphological traits. However, abruptly in 1933-1937, the archeologists Egorov and Lunin (1937) discovered that Rengarten had made a mistake: the bones dated from historical times, specifically from the Bronze Age. Isn't Rengarten's error worth immortalizing?"

However, that proof of the survival of Neanderthals well beyond the Paleolithic and even of the Neolithic ages did not convince those anthropologists blindly attached to the dogma of the ancestral ape. Faced with the fact that there was no doubt about the dating of the Podkumok skullcap, they claimed that it actually belonged to a *Homo sapiens* who had preserved some Neanderthaloid traits, and thus an individual going through a belated transition between the Neanderthal to the Sapiens stages.

By the way, since it is well established that in Western Europe pre-Sapiens "differing in no essential trait from today's type" came before the first Neanderthals, and that the latter, of a generalized type close to the Sapiens, were later followed by more extreme types, how is it possible to believe, as even Porshnev does, that Neanderthals then gave birth to the form from which they clearly derive?

Other remains of skulls and skeletons similar to the Podkumok skullcap have been found in various areas of Central Eurasia, particularly at Khvalynsk on the Volga, and on Undory Island. Following the deplorable tendency to base a diagnostic on the age of the terrain rather than on the morphology of the specimen, the German anthropologist H. Ullrich first wrote in 1958, rather prudently, that those two pieces could not be dated (*night datierbar)* but, more significantly, declared that "Although they exhibit some undoubtedly Neanderthaloid traits, they most likely date at the earliest from the upper Paleolithic and then *must* be attributed to *Homo sapiens diluvialis.*" (I italicize *must* myself to make the point.)

Even more recent than those remains from the upper Paleolithic, which must have lasted from -35,000 to -10,000, are those from Kebeliai, discovered in 1950 in a gravel pit near that Lithuanian village of the Klapeida region. They consist of a frontal bone and of small parietal fragments. In spite of their clearly Neanderthal characteristics (heavy eyebrows, narrow receding forehead), Goudelis and Pavilonis estimated, in 1952, that they must belong to a modern man. And that's because these bones were too recent, dating from the period named after the "Mer des Littorines," which corresponds to the end of the Mesolithic or the beginning of the Neolithic. That skullcap, at least Neanderthaloid, if not Neanderthal proper, would not be older than 5,000 to 6,000 years.

So, each time Neanderthal remains are found in deposits that are deemed too recent, it is decided that the specimen was older than suspected, or that it could not be properly dated, or that it was not a real Neanderthal but some *Homo sapiens* having preserved transitory Neanderthal features.

Wouldn't the simplest, most logical, and most obvious and honest conclusion have been to admit in each case that Neanderthals had survived much longer than expected? Since one must recognize that Neanderthals and Sapiens co-existed on Earth from the Mindel-Riss interglacial (Steinheim pre-Neanderthal, Swanscomb pre-Sapiens) until the end of the early Würmian, for a period of over 400,000 years, there is nothing that prevents thinking that such a co-existence, most likely not peaceful, could not have lasted another 35,000 years. One can certainly understand that it could have lasted until the late Würmian, judging from the remains of Khvalynsk and Undory, until the

Neolithic, as evidenced by the Kebeliai skullcap, until the northern European Bronze Age, as shown by the Podkumuk skull, and even until the Iron Age, as suggested by the Novosiolka skull. Frederic Zeuner, one of the world's greatest experts in the matter of dating fossils, was aware of that possibility when he wrote in 1958: "It is quite possible that Neanderthal Man might have survived here and there during a later (than the Würmian) period, but there is not enough proof. Conditions in Russia must be studied further."

If Neanderthals have really survived through prehistory into historical times, they should have left traces of their existence in human chronicles through the ages. We already know from Porshnev's work that this is indeed what has happened; they have been the subject of a profuse literature from the remotest antiquity until this day. But how about before, before the invention of writing, were they never pictured by human artists? If Sapiens and Neanderthals developed in parallel during hundreds of millennia, it would be most surprising if the former had never thought of drawing the latter, as they did for most of the larger animals they lived with.

I hasten to point out that the reason I will show some possible and potentially debatable examples of such representations is mainly to refute such a potential objection and not to bring any further proof of the survival of Neanderthals beyond the early Würmian. Such representations are merely to be seen as complementary information on the issue, at best an additional *a posteriori* contribution. In any case, the search for prehistoric representations of Neanderthals could not have been undertaken before we had learned more about their external appearance, or at least that of their ultimate descendants.

As far back as the Aurignacian period, the very time when Neanderthals were traditionally thought to have disappeared, there was in Western Europe a great artistic blooming that manifested itself through numerous etched, painted, or sculpted representations of animals and human beings. The latter are rather rare and surprisingly mediocre for such apparently gifted artists. Boule and Vallois remark that "On the whole, the anthropomorphic representations of the quaternary age strikingly contrast in their clumsiness and inaccuracy with the many animal figures, many of which are masterful pieces of art." These authors go as far as speaking of "the inability of artists of the Reindeer age to draw the human body."

However, an artist capable of sketching with much refinement and realism a mammoth, a wooly rhino, a reindeer, a horse, a bison, or a bear, cannot be incapable of sketching one of his own. His apparent clumsiness is thus probably deliberate; in a world dominated by magic equivalence, no man would accept to be so represented. He would be much too concerned that someone could harm him by doing something to his image. For example, in drawing arrows piercing it.

That's why human bodies are usually reduced, in hunting scenes, for example, to stick figures. Even in those representations whose function should have required somewhat more realism, as in those steatopygic Venuses, the pin-ups of the Paleolithic, the face is covered by hair or simply not shown. In some more detailed sketches of men, they are wearing masks. Finally, when faces are drawn as in the Marsoulas Cave (Haute Garonne department), they are mere coarse caricatures, or rather since caricatures are simplified in order to emphasize some resemblance, schematic sketches similar to kindergarten drawings.

Realistic representations in the Isturitz cave (Basses Pyrenees Department) of what appears to be a Neanderthal. (A) For comparison, schematic representations of a human from the same cave (B) or in Marsoulas cave (Garonne Department) (C and D). (After Mauduit, 1954)

That being the case, one should be somewhat surprised to find some human representations that are just as detailed and carefully executed as animal pictures: specifically, the "Reindeer Woman" of Laugerie-Basse in the Dordogne area (whose head is unfortunately missing because the wooden support is broken), the couple from the Isturitz Cave (Basses Pyrenees Department), and the "hairy" man from the same cave.

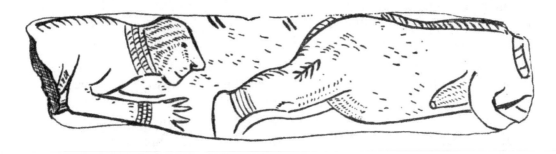

The two sides of a Magdalenian engraved bone from Isturitz, showing on one side a pair of bison, and on the other a pair of "hairies" (After De Saint Pierre).

But the reason for these exceptions is obvious: these are not representations of co-specifics, but of creatures considered as enemies, or even more so as preys, mere animals, unable to get back at you through magic.

The fact is that contrary to all the other human figures depicted in Franco-Cantabrian cave art, all of them are entirely covered with hair and drawn as carefully as the animals shown near them. Perhaps even more revealing, the females are not steatopygic.

Let's look more closely at the bone engraved on one side with a couple of hairy humans, and on the other a pair of bisons. Boule and Vallois wrote about it: "Below the woman we see the upper part of a man's body adorned with necklace and bracelets. The head is *obviously a caricature* [my emphasis]. The arms are bent and raised, as are those of the woman, in an attitude of prayer or supplication. The arrow seems to indicate that this is art with a magical purpose, perhaps expressing a sentiment of amorous lust." Similarly, in his book *Quarante mille ans d'art moderne (Forty Thousand Years of Modern Art)*, Jacques Mauduit wrote: "The woman is stout and hairy. On her thigh, there is an arrow with a triple row of barbs, a symbol of conquest.

The engraving on the other side of the bone is not without relation to the former one. It shows a male bison ready to mount a female whose hindquarters and raised tail are visible; the male also has a pair of triple barbed arrows on its shoulder."

There seem to be a number of contradictions between these two commentaries. Even if we were to admit that the symbolism of Cupid's arrow goes all way back to the Magdalenian, which remains to be proven, why is it that the barbed arrow is shown on the woman's thigh in the human couple and in two copies on the bison's shoulder? It doesn't make sense for bovines to consider the male as the amorous object of the female. Even if one were to suppose, given the fragmentary nature of the bone, that there was an arrow on the buttock of the hairy man and another on the flank of the female bison, it would seem extravagant to conclude that there was a reciprocal "amorous desire." Besides, if the hairy man is really in a begging attitude towards the woman, who might she be addressing a similar solicitation when her back is turned to the male? If they were really addressing the same prayer to each other, that would make no sense at all; there is no need to beg for what is offered! Why go through such a rigmarole if they both agree?

Isn't the simplest and more logical explanation that this carved bone foresees or commemorates a hunting scene, or perhaps its desired outcome? I favor the latter option because of the presence on the edge of the knife of two sets of double notches traditionally known as "hunting marks." Such marks are used today; western lawmakers, big-game hunters, and other killers sometimes scratch notches on the handle of their gun; military pilots draw the silhouettes of downed enemies on their fuselage. As Goury clearly explained, such marks are "undoubtedly mnemonic devices used to keep alive the memory of some events or to keep track of scores."

In view of all this, it's likely that the etched Itsturitz bone simply means: "I shot a couple of bison and a couple of hairy ones."

One might even ask, purely speculatively, if what one imagines as a necklace and a bracelet isn't rather the rope used to tie up the two creatures after having put them down with an arrow. Their "begging or praying stance" would simply be that of people whose hands have been tied together.

If the two hairy ones had merely been considered prey, there is no reason not to represent the male's head with as much realism as that of the bison. One can recognize in this man the strongly receding forehead, the short neck and the strong nape, the weak chin and especially the upturned nose of our specimen and of its Asian conspecifics: extreme Neanderthals.

The resemblance is even more striking when looking at the more detailed portrait of the bearded head from Isturitz. In contrast to the quasi-abstract and at best schematic representations of the Sapiens of the time, his hair is shown with great attention to detail: the face is surrounded by hair, visible as far as the back of the neck; the hair is longer on the head; and, perhaps stretching the evidence, the eyebrows are sparse and the hair on the cheeks is scattered. In any case, the upturned nose is unmistakable.

If no one has ever thought of these various sketches as representations of Neanderthals, it's partly because their external appearance was still unknown, and also because they *date from a time supposed to be later than their presumed extinction.*

Comparing the portraits of Paleolithic Neanderthals with those of today suggests that their appearance has not changed markedly over the past tens of millennia. But how about their behavior and lifestyle?

While the descriptions of pongoid men might allow us to fill some gaps in our knowledge of the external appearance of classic Neanderthals, might our knowledge of the biology of the Asian wild men inform us on the lifestyle of their ancestors from the early Würmian? That is indeed likely for those traits closely linked to their physiology, such as the mating season, the length of the gestation period, the type of diet, the adaptation to low temperatures, mountain climbing, and evening or nocturnal life,[57] as well as the development of the sense of smell and the range of vocalizations. But that is most unlikely when considering their cultural activities. There are striking differences between what we know about fossil Neanderthals and pongoid Neanderthals, relating particularly to tool-making, the use of fire and the power of speech, with consequences for their social life, the building of shelters, and the intellectual development accompanying make-up and jewelry, funeral rites, and perhaps a cult focusing on some animals.

Classic Neanderthals are associated with Mousterian stone tools. Their use of fire during the Würmian glaciations is documented by the discovery of relics of hearths, pockets of ashes, in a number of Mousterian sites. In the Basura cave, in Toirano, where a number of unmistakably Neanderthal footprints have been found, a number of charred twigs have been found, probably the remains of torches. The topography of some Mousterian sites suggests that their occupants lived in tents or in round huts dispersed over more than 50 hectares (Trecassats site), basically a village. Perhaps sometimes they lived in a large shelter greater than 80 m^2 [860 sq. ft] home to more than one family (in the Vaucluse and Gard Departments). Thus, the makers of the stone tools found on the site were members of rather large bands and must have had some degree of social organization.

It's also pretty certain that some old-time Neanderthals used red ochre as make-up, that they wore bone or shell pendants as well as perhaps semi-precious stones.

A number of fossil Neanderthal skeletons have been found that were clearly

---

57   One of the striking features of the classical Neanderthal skull is the size of the orbits, to which naturally correspond large eyes, best adapted to crepuscular vision. That is indeed the case among nocturnal primates, lemurs and especially tarsiers, which have extremely large eyes, and also in the South American douroucouli, the only exclusively nocturnal monkey. The fact that the extreme Neanderthals could well have been mostly nocturnal supports the idea that they were the ancestors of today's wild men.

buried, as at Regourdou, Moustier, and La Ferrassie, in the Dordogne Department. Near some tombs, animal bones were found, which appeared to be offerings to the dead; that is the case at Techik-Tach, in Uzbekistan.

Finally, in some burials rather unequivocal traces have been found of a bear cult (Le Regourdou, Drachenloch).

How is one to reconcile these clear signs of a relatively advanced culture in the past with the present state, a strictly animal life, of relic Neanderthals? As I have said already, it is that sharp contrast that has long kept me from fully accepting Porshnev's thesis, until the day when the study of a specimen forced me to change my mind. When faced with facts, opinions must give way. It is opinions that must be modified or even abandoned; facts are not to be ignored or hidden, as they have too often been in anthropology.

What bothered me most was that according to unanimous reports by eyewitnesses, the wild hairy men sometimes used stones or sticks but never fabricated tools. However, all agree that even the inferior hominids like Sinanthropes (Peking Man), created stone tools. And after a brilliant demonstration in Johannesburg by James W. Kitching, the main collaborator of Raymond Dart, I cannot doubt for a moment the reality of the osteo-dento-keratic tool making of the Australopithecines: weapons and utensils made of pieces of bones, teeth, and horns. How can one imagine human beings anatomically similar to us but with intellectual development of anthropoid apes, incapable of creating the simplest tool beyond using a branch?

These facts require explanation. *A priori*, three come to mind: (1) Neanderthals never made any tools; (2) they have lost the skills; (3) some did make tools, others didn't, and these latter ones are those who survived.

As shocking as it may seem, that first possible explanation cannot be merely shrugged away. In the realm of prehistory, many interpretations and conclusions are extremely fragile. For example, one is not quite as sure as one had been previously if the paleontological concept of Neanderthal is necessarily associated with the archaeological idea of Mousterian. Remains of Sapiens linked with Mousterian artifacts have been found, specifically at Djebel Qazfeh, in Palestine. Conversely, isn't it possible that some Neanderthals would not have been toolmakers?

Could one perhaps go as far as imagining, as was freely done for Sinanthropes and Australopithecines, that the skulls and other Neanderthal bones found in Mousterian sites were merely kitchen scraps from contemporary Sapiens? In support of this idea, we could mention the frequent dispersal of Neanderthal remains (La Quina, Montsempron, etc.) and the unmistakable evidence of carving or breaking up of their bones (Krapina, in Croatia). As to the ceremonial inhumation of Neanderthals, might we perhaps imagine, in the light of what we know about their descendants, that such ceremonies were in homage to beings held as sacred, or as faithful servants, perhaps even particularly cherished domestic animals? After all, today we have cemeteries for dogs and cats. In that case, the Neanderthals' jewelry might be compared to that which adorns idols or to the fancy collars worn by favorite pets. Finally, the bones of bears found in some Neanderthal burials would simply be due to an association in the mind of Sapiens of two species, similar to the moniker "bear-man" often given to their descendants today. It would have been to honor and distinguish them that the graves of hairy men would have been crowned by bear skulls, much in the way that today one puts a cross over a Christian tomb or a Star of David over a Jewish one. One might even let one's imagination wander further…What if a tomb containing all kinds of bear and Neanderthal bones was only the storage area of some primitive pharmacist, a shaman responsible for the preparation of the real *moumieu* or, for lack thereof, its closest imitation, bear grease.

You may think these interpretations are rather far-fetched. Perhaps, but believe me, they are no more so than many current hypotheses in prehistory. However, they face some serious objections.

During the second half of the Mousterian period, which coincides with the early Würmian glaciations and precisely the same period as the Neanderthal burials, not a single bone clearly belonging to a Sapiens was found in Europe. Of course, the representatives of our species, who were already present in Europe during the previous interglacial periods, had to be somewhere during those times. They were, but apparently not in those areas that were too cold and where only the Neanderthals could live… and die. And anyway, if the Sapiens had wished to bury their devils, their slaves, or their pets, they would surely have buried their own dead. In that case, their tombs would have been found next to those of the Neanderthals.

Since that is not the case, one may safely reject the explanation according to which the paleanthropes had *never* made any stone tools, and that *all* such tools were due to contemporary Sapiens.

The second explanation, which says that the Neanderthals had gradually abandoned all stone tool-making, is that favored by Porshnev, who attributes it mainly to the gradual thinning of the great herds of herbivores. This process is most pertinent since it is known to have often taken place in human history. Ernst Mayr, a prominent analyst of the mechanisms of human evolution and speciation, wrote in 1963: "The history of peoples and tribes is full of incidences of a secondary cultural deterioration, *vide* the Mayas and their modern descendants! Most of the modern native populations with rudimentary material cultures (e.g. certain New Guinea mountain natives) are almost surely the descendants of culturally more advanced ancestors. This must be kept in mind when paleolithic cultures from Africa and western Eurasia are compared with those of southern and eastern Asia. Stone tools and the hunting of large mammals seem to be closely correlated. Could such peoples have lost their tool cultures after they had emigrated into areas poor in large game? Could this be the reason for the absence of stone tools in Javan *Homo erectus*?"

So, a cultural decadence could be due as much to a *local* disappearance of large mammals as to a gradual departure from areas where they roam. Did Neanderthals have some reason to retreat southward? Yes, of course: the rivalry and competition of the Sapiens, their contemporaries, and of the wars that took place every time they met.

All over the animal kingdom, it's the closest species that engage in the fiercest competition and try to eliminate the other; they have essentially the same aspirations, the same desires. Given their very similar anatomy, the two most closely related forms of prehistoric men must have vied through violence and wile to occupy the same ecological niche.

As they were much better adapted to the cold than the Sapiens, Neanderthals first had the upper hand whenever the glaciers advanced. The Sapiens then had no other choice than to retreat towards warmer climates or perish on the spot. However, during each interglacial, they came back, armed with ever better stone tools, while the Neanderthals who had remained in place, curiously enough, showed no technical progress. Was it the fact that since they had not developed speech they had to rely on learning strictly by imitation? We shall return to this point. Anyway, for a period of 70,000 to 80,000 years, Mousterian tools showed practically no change. So, while at the beginning, the two species were fighting on nearly equal terms, the Sapiens soon took the lead. When, after the second phase of the Würm glaciation, the Sapiens returned with better weaponry *and with the means of surviving the cold weather,* it was game over for the Neanderthals. From being rivals, first feared, then defeated, they became poor cousins, parasites, prey, and perhaps even domestic animals. To survive, all they could do was to take refuge in the most inhospitable or uncomfortable regions of the Earth. Not only was there a lack of big game in such areas, which rendered their stone weapons unnecessary, but also perhaps they no longer could find flint stone or other kinds of materials needed to make them.

Thus, *the Neanderthals would have degenerated technically not only because they had changed environment, but also because they had been forced to do so.*

Nevertheless, even with the addition of such nuance, Porshnev's explanation doesn't quite satisfy and remains insufficient. In Western Europe, during the early Würmian, Neanderthals created lanceolate biface tools, built large shelters, decorated their bodies in various ways, lived in organized groups, buried their dead with offerings for the afterlife, and worshipped the Bear. It is difficult to imagine that creatures possessing a seed of technical, esthetic, social, and religious sense would simply retrogress to adapt to its environment just like a mere animal. The fall would be too hard, the regression too dizzying! Whatever one may think, evolution is to a certain degree irreversible, for the simple reason, astutely stated by Heraclitus: "One never bathes in the same water." So, one may finally wonder—and that's the third explanation—whether all Neanderthals had reached the same cultural level.

It would seem ridiculous to expect a uniform level of cultural development within a population dispersed over an enormous continent. We need only look around us to see that today, on the Earth, all levels of technical development are found, from hunting-gathering, using stone, wood bone, or composite tools, to the level reached by nuclear physicists or space scientists. Why wouldn't such a similar diversity have been found among Neanderthals?

One might point out that today's differences are due to the extraordinary acceleration of technical progress since the end of the late Mousterian in Europe, namely over the past 35,000 and especially over the last 5,000 years, since the first use of melted copper and the invention of writing to the taming of nuclear energy. It is obvious that in a race the differences in arrival times are the greatest when a greater diversity of means of travel are used: on an extended course, those differences will be much greater than in a simple foot race if the means used range from the bicycle to the jet airplane. And that is indeed what we observe: among our Neanderthals, corresponding to simple foot runners, there is not that much difference between those who, like chimpanzees, throw stones at their enemies or hit them with sticks, and those who routinely make a few flint tools. These differences are in no way comparable to the abyss that separates Australian aboriginals from American or Russian astronauts.

What makes a difference in technical levels among the various Neanderthals once dispersed all over temperate Eurasia more likely is the polytypic nature of their populations.

I am not speaking here of the differences between the classic Neanderthals and the pre-Neanderthals. These generalized types only represent an early evolutionary stage with respect to the well-defined Neanderthals of the Würmian period. But at the same geological level, during the Riss-Würm interglacial for example, the European pre-Neanderthals are clearly different from those of the Middle East. Not just because the latter began to be characterized earlier, but also because they evolved in slightly different directions. They differ from the former by having a divided brow-ridge. Among the classic Neanderthals of Europe, the ridge above the eyes forms a single arch, while among those of western Asia there is an individual swelling over each orbit, forming a split ridge.[58]

During the following geological period, the early Würmian, classic Neanderthals from Europe and Asia are even more different. Those from the East are even more

---

58   Le Gros Clark (1955) calls this feature *arcus superciliares*, in contrast with the *torus supra-orbitalis* proper. It is interesting to find out that this feature also allows one to recognize the geographical origin of old-world monkeys. I noticed that it is often easy, using that criterion, to distinguish at a glance an African from an Asiatic monkey, whether they be cynomorphs (dog-like apes) like semnopithecus (langurs), or colobus monkeys, or anthropomorphs, like the orangutan or the gorilla. The difference is particularly striking in adult males. That same criterion confirms that the only species found in both Africa and Asia, the macaque, is probably of African origin and an invader in much of Asia.

specialized (for example the Kiik-Koba male) than the more "extreme" ones of the West.

Thus there were among fossil Neanderthals great geographical races, real sub-species. And that's not all. One also finds among them, even at the same place and time, important individual variations. For example, it is significant that for the two adults of La Ferrassie (Dordogne department) the carpal-metacarpal articulation of the thumb is in the shape of a saddle, as it is in Sapiens, while that is not so in other Neanderthals of the same region (for example that of La Chapelle aux Saints). John Napier, who is an expert in the matter, thinks that this type of articulation is the essential sign of the opposability of the thumb and thus an essential condition for a fine precision grip, the ability to grasp an object very precisely between thumb and forefinger (perhaps with additional help from other fingers). The presence of such a structure has a determining influence on manual dexterity and an individual's technical abilities.

Everything thus seems to point out that depending on the region and the individual, Neanderthals had reached very different levels of anatomical, intellectual, and cultural development.[59] Some perhaps never got out of a strictly animal-like existence. Only the most gifted, or specific family groups or tribes, might have evolved to the point that they could fabricate more or less primitive stone implements, with all what that might mean for their lifestyle.

Finally, the answer to the technical decadence of Neanderthals might be found somewhere between the last two explanations put forward, or perhaps as a subtle mixture of both. Pongoid paleanthropes without any culture who survive today are most likely the descendants of Neanderthal populations whose skills have deteriorated, and of individual Neanderthals who never learned to make stone tools to improve their life and have survived mainly because of their more animal-like nature.

In the days when Sapiens and Neanderthals were competing for the same territory, the most skillful and best armed among the latter must have been regarded as the most dangerous rivals of our species, and therefore needed to be eliminated first. They were then probably the first to disappear. Those who were not tempted by the "human adventure," either because they lacked the potential or were too lazy, remained closer to their animal nature—wild beasts always on the look-out, experts at running away or hiding—and had better chances of escaping the massacre. Not only because they were seen as less of a threat to Sapiens, but because they could more easily escape persecution. Indeed, while the Sapiens were clearly superior to the Neanderthals through their manual dexterity and skill at developing new weapons, the latter were superior to the humans in many other aspects.

They outdid humans in their adaptation to cold temperatures, their agility in mountainous terrain, and their better night vision. The nomadic and solitary life

---

59   On a basis of an extraordinarily careful study, Professor François Bordes managed to bring out a certain cultural difference among Neanderthals within the French territory alone.

was also easier for them because of their keener sense of smell and their stronger vocalizations.

Because of their hairiness, their highly developed sinuses, and their strong and stocky build, Neanderthals could, if they wanted, distance themselves from Sapiens by moving to colder regions, further north. But it was also possible for them to remain nearby, in the hope of becoming parasites of their rivals, by climbing up nearby mountains, where their hooked feet and long, even-fingered hands worked marvels at grasping the smallest projections, and where no one could catch them. An observer of one of their descendants has compared their progress on a steep cliff to that of a spider in its web. On that topic, some paleontologists have pointed out that *Neanderthals preferred hills and plateaus, while Sapiens lived mostly in valleys and plains.*

To judge from the biology of their descendants, it would appear that Neanderthals had adapted physiologically to life in mountains and ended up behaving just like their neighbors, the bears, with whom they already had so much in common (hair, flat feet, an omnivorous diet, life in caves). When winter came and starvation threatened, they would eat more than usual and become fat and chubby, accumulating fat reserves that they could assimilate slowly, which would allow them to sink during the colder months into a deep sleep, a semi-hibernation. I imagine that this remarkable ability to go without eating would always have been a matter of admiration and envy by Sapiens, who were also under-fed and sometimes perished during the cold season. It might have led them to believe that the fat of hairy men had more than just nutritious value, but a rejuvenating power. Just as sexually impotent Orientals imagine that they will find solace in the horn of the rhinoceros, faint and weakened Orientals think that they will find the secret of rejuvenation in Neanderthal fat, which allows one to live without eating. Originally mere nuisances, the poor hairy savages have become a sought-after game, their fat transformed into *moumieu*. Another good reason to stay away from the Sapiens!

With their bigger eyes, Neanderthals had the option of vanishing into the night, as lemurs had done to escape competition with monkeys, hiding during the day in holes and bushes and becoming active at dusk. It goes without saying that their superior sense of smell, enhanced by their large forward-looking nostrils, was of great help in their nocturnal wanderings and helped them maintain a prudent distance from potential danger.

Since it is much easier to find and to hunt a herd of animals than solitary game, groups of defeated Neanderthals must have scattered like the remains of a routed army, with individuals wandering randomly. Perhaps they were not very gregarious to start with. In any case, once dispersed they would have no problem finding each other when they wished. With their acute sense of smell, they could easily detect the fetid odor they emitted and which Sapiens found disgusting. But it was mainly their fantastic vocalizations that allowed them to announce their presence far and wide to any interested parties.

In order to avoid contact with Sapiens, the paleanthropians had to flee into the cold, into the hills, into sleep, into the darkness and into solitude, to flee perpetually. FLEE. But flee on the spot, becoming for their triumphant rivals *invisible and elusive,* while still benefiting from the vicinity of the Sapiens: cleaning up their largest preys, along with vultures and coyotes, pilfering from vegetable gardens, eating leftovers from meals and kitchens. That was the best way of ensuring both their safety and easy access to additional food supplies.

There is a flip side to everything. There were also disadvantages to this flight in all directions, which ended up for the most developed Neanderthals to a regression on the road to social and cultural progress.

By choosing, for the sake of safety, to dissolve their groupings, and to live only in couples, and perhaps so only in the mating season, the Neanderthals gradually lost whatever social organization they might have had during the early Würmian. They abandoned all social life, which requires a variety of communication skills and stimulates the growth of an ever more complex language; they also abandoned working together, which favors the improvement of tools.

Such tools were anyway bound to diminish in abundance and eventually to disappear. It's possible that in the new areas where they took refuge, there was less likelihood of finding pebbles or breakable stones with which to make their tools. But, mainly, there is no point in sweating over the making of stone weapons when there is no longer any question of competing with Sapiens who dominated the savannas and the sparse forests where the big game lives. And besides, now that they had become solitary or nearly so, how would the hairy men have managed to attack large animals, which could be hunted only in groups by herding them towards cliffs or specially dug trenches?

And since fire is only an accidental by-product of the making of flint tools, as Porshnev so cleverly demonstrated, its use was also lost among devolving Neanderthals.

By opting for a nomadic lifestyle, they also abandoned the construction of permanent dwellings. They probably first made some tents, easily set up or put away, but finally ended up discarding them as too conspicuous, finding shelter only in caves or holes in the ground, or bedding hidden in thickets. Besides, to make tents, one needs large skins and to get them, one has to kill large animals…. It all fits together.

And as to painting oneself with ochre and wearing jewels, that was no longer feasible after some time wandering in the wild. Survival monopolized all aspects of life.

Forever on the run, Neanderthals of course also stopped burying their dead. They had neither the time to dig adequate tombs nor the ability to do so without stone tools.

Of course there was no question of leaving offerings or adorning tombs since they no longer had any possessions. On the other hand, for those who never had access to meat in quantity, there was a great temptation to simply eat up their dead, which they probably did. It all fits.

How about the cult of the Bear, about which much has been said. It was probably no more than a simple primitive totem. In their mountain retreats, Neanderthals often ended up living in caves where bears also lived. They had to fight bears for the same shelters (which they couldn't do any more once they lacked weapons) or to share these shelters with them. Hairy as they were, the paleanthropes must have felt closer kinship to bears than to the "naked apes" who lived in the plains and valleys and hunted them without mercy. Neanderthals ended up imitating the life style of the large plantigrades, even imitating their long winter slumber so as to be able, like them, to get through winter, the "killer of poor folk." In short, in their own eyes as well as in those of the Sapiens, they were the bear-people, which they have remained to this day.[60] It's not surprising that they would have expressed a degree of respect, based at once on fear and admiration, towards bears, whose skulls they preserved as precious relics. Such a feeling might not have been so different from that experienced by baboons and gorillas for the alpha-male head of their family or tribe. If that is religion, it's at best a sketchy one.

Everything hangs together after all. One downfall precipitates another, and those Neanderthals who were evolving towards humanity were either exterminated by conquering triumphant Sapiens or forced to turn back and return to their animal-like lifestyle, after losing along the way all that they had acquired in terms of technology, social organization, religious sentiments, and creativity.

The study of wild children, wolf-kids as they are sometimes called, clearly demonstrates the incredible regression the most civilized *Homo sapiens* can experience when returned to nature. However, the main reason for this regression is not the wildness of its situation as much as the fact that the child finds himself precociously *alone*, isolated, without any means of acquiring through its parents or some other adult the cultural heritage of its ancestors. Not only is he unable to speak, but he will never learn to if he lives alone for the first few years of his life.[61]

Having been raised without the sounds and gestures of his parents, as are the young of animals, it sometimes happens that the wild abandoned child does not even learn some human traits as basic as a bipedal stance, and that it is often hard to teach him to adopt it. In this way, he stands below an animal that has normally been raised by its parents or older relatives and learned by imitation.

---

60   As one may conclude from the name often given to wild men: *jen hsung* (Chinese), *mi-dre* or *mi-teh* (Tibetan), *iou-woun* (Burmese), *michka-tchelovek* (Russian), etc.

61   Cf. Malson, 1964.

Within its family unit, the young devolved Neanderthal was certainly raised by his mother, or perhaps by both parents. But how? By words, as young Sapiens, or by noises, gestures, and mimicry as with all other animals? Did Neanderthals possess an articulated language and conceptual thoughts? To speak or not to speak, that is the question. If they did not speak, it would be even easier to explain their rapid deterioration and the regression of their culture.

Whether Neanderthals were capable of speech or not has been a matter of lively controversy, and we cannot enter here into details; that would carry us too far. Let's just briefly survey the issue.

It was once thought, and some still believe this, that it would be possible to ascertain whether some hominid was capable of speech from its skeletal remains. For example, it was long claimed that the presence on the mandible of the apophyses, where the muscles of the tongue are attached, is a sign of articulated speech. However, many modern people lack these small bony protrusions and can still speak. Some researchers have not given up and continue to base their arguments for the lack of speech in Neanderthals on other skeletal structures.[62]

However, most of the arguments aimed at resolving this question have been based on the form and structure of the brain, and thus for fossil men on endocranial casts. However, already in 1915 Symington showed that the latter are not impressions of the brain itself and cannot be used to map its shape. Nevertheless, some Soviet researchers have claimed that the relative development of some cerebral areas is sufficiently reflected in endocranial casts to state that speech is the monopoly of *Homo sapiens*.

A number of authors would rather base their conclusions on this matter on the external manifestations of intelligence. They take into account the quality or the variety of artifacts, as well as such practices as the burial of the dead, accompanied by offerings. In their view, belief in an afterlife implies the need for symbolic thinking and a complex language to communicate that belief.

A careful analysis of Paleolithic tools by Professor V.V. Bounak reveals that their manufacture was an extremely stereotypical and nearly automatic process. During the early Paleolithic period, any change in pattern corresponded to about 1,000 generations (of thirty years); in the mid-Paleolithic, at least 100-200 generations. Such long stages undoubtedly don't result from conscious control and are based on deeper processes. Speech thus didn't play any role in the transmission of technical progress at that time: only imitation was involved.

One should then admit that the extremely long period (70,000 to 80,000 years)

---

62   Of particular relevance to this issue are the acoustic experiments of two American scientists, Lieberman and Crelin, who in 1971 managed by reconstructing the pharynx and the larynx of the La Chapelle Man to prove that European classic Neanderthals were incapable of pronouncing many vowels and consonants, and possessed at best one tenth of the speaking ability of modern man. Lieberman's further studies only confirmed these results.

over which Mousterian tools showed practically no change suggests a Neanderthal technology based on mere imitation.

At the time when the still undifferentiated population of pre-Sapiens from which evolved both modern Sapiens and Neanderthals began to split, most likely under the influence of climatic factors (probably towards the end of the Mindel-Riss interglacial), each population went its own way, taking with it its different genetic baggage relating to manual dexterity and predisposition to articulated language. For lack of being able to communicate orally to their young ones any technical refinements discovered either by reasoning or happenstance, the Neanderthals remained for tens of millennia frozen in an unchanging methodology. That knowledge was most likely easily and quickly lost when their lifestyle was perturbed by the assaults of the Sapiens, who were perfecting their own technology. Just as there is nothing more detrimental than over-specialization when external conditions change, nothing is more fragile than stereotypical behavior when it can no longer be followed exactly.

One should nevertheless respond to the objection that some Neanderthal customs indicated a high level of development of conceptual thought and required verbal communication. I have already disposed of the argument based on the Cult of the Bear. What about the burial of their dead, accompanied by funeral offerings?

Some pre-historians have greatly exaggerated the importance of religious thought among fossil humans. The fact that some would choose to see a "Cult of Fertility" in images that are most closely related to bathroom graffiti provides a measure of their errors.

To attribute a religious motive to the burial of a corpse is equivalent to attributing a knowledge of hygiene to a cat that buries its excrements. Eating the brain of an enemy doesn't necessarily indicate a wish to assimilate its soul; it might simply reflect a gastronomic taste for brains. How in the world would pre-historic men have known that the brain was the seat of the psyche, while over many centuries of our history even the most erudite have usually placed it in the heart or even the liver?

The inclusion of some objects in a tomb, especially objects having belonged to the deceased, clearly implies some reasoning (what long went together should stay together), but such reasoning, far from being a syllogism, does not surpass, even in complexity, that which is rightly attributed to monkeys following far subtler behavior linked to abstract concepts (for example, accumulating tokens for future use in an automatic candy dispenser, or stating a precise request using deaf-mute sign language, or the grouping of symbolic drawings).[63] After all, to provide the dead with some provisions, as one does to a traveler, might be interpreted another way, as a sign of deep stupidity, and certainly not as a sign of rational thought. One should dare to view

---

63   Such experiments were performed on chimpanzees in the United States by Dr. John Wolfe, of the Yale University Primate Laboratory; by Drs. R. Allen and Beatrice T. Gardner of the University of Nevada; and by Dr. David Premack of Santa Barbara University, in California.

the question from the point of view of Sirius, as if one was analyzing the behavior of members of another species… insects, for example. Is the belief in an afterlife a proof of intellectual superiority?

A magical belief based on analogies—often very superficial resemblances, such as between death and departure for a trip—is certainly a sign of symbolic thought, but certainly not of rational thinking.

Such a mode of thinking based on similarities, things imitating one another, is merely an association of appearances that is naturally obvious to everyone, and doesn't require communication between individuals through a conceptual language. A human brain looks like a walnut and a mushroom like a phallus and words are not required to explain it. In their simplest manifestations, beliefs and magical rituals don't need transmission by tradition. Besides, the simplest rituals can readily be communicated by imitation.

It is thus quite legitimate to think that Neanderthals easily reached their highest cultural level without the need for articulated speech.

Porshnev is quite right to consider that a Primate transformed into a Man (in the strict sense of the word: a *Homo sapiens*) only when he acquired the power of speech. Furthermore, the use of speech could only become influential once people became gregarious. Only then could the accumulation, by working together, of the knowledge of even the smallest technical improvements be communicated to the entire tribe and even to other tribes in exchange of their knowledge. This process snowballed into the fantastic, even explosive, acceleration of technical progress that now characterizes humanity.

The vertical posture, which favors the full development of the brain at the top of the spine, and the bipedal stance, which frees the hands from a function of support and locomotion, followed by the refinement of those organs, were necessary factors in the development of man, but only in a preparatory sense. It's only thanks to the complex development of the cerebral cortex, allowing conceptual thought, tightly linked with the power of language, and to the manufacture of ever more complex tools thanks to verbal communication of technical skills, that Modern Man came to be. Language, work, and society are the three decisive factors in hominization.

These facts confirm the soundness of Engels' theory of human genesis through work, a corollary of a fundamental thesis of Marxism. This also explains the fundamental importance for Porshnev and his fellow Soviet researchers of the relic Neanderthals, creatures anatomically similar to Man, but without articulated speech, social organization and industry.

We note, however, that while Man may be the result of the use of speech by a population of bipedal, social, tool-making primates, that process did not occur slowly and gradually, at least on a geological scale, as the science of anthropogenesis would long have led us to believe. Hominization took place rather abruptly within an already well-differentiated lineage… an idea that fits very well with the traditional spiritualist thesis, which says that at a certain time, a "soul" was given to Man and only to Man.

This is worth mentioning, since it is very rare that the philosophical implications of a scientific discovery turn out to satisfy materialist thinkers while also reassuring religious thinkers.

Chapter 12

CAIN vs ABEL
De-hominization: a new perspective on the origin of man

"One should not expect great progress in Science by
continuously working on the same old overloaded
main trunk. Everything must be renewed, down to
the deepest roots."— Francis Bacon

"What I like in Man is the Ape, not the Angel…"
— Sir Robert Sitwell "Aspiring Ape"

According to an old Jewish legend, the members of one of the three groups of men who built the Tower of Babel were transformed into apes in punishment for their impiety. Many tribes in Africa, Indonesia, and even South America firmly believe that apes are fallen men, perfectly capable of speech but not willing to do so for fear of being put to work. In his famous *Description of Africa*, published in French in 1668, Dr. Olfert Dapper wrote that the *quojasmorrou* (probably chimpanzees): "They are descended from men, say the Negroes, but they became half-beasts by always remaining in the forest."

As one can see, the concept of "de-hominization," which fits so well the particular evolution of Neanderthals, is nothing new. It is at least as old as the hoary fable of "humanization" of the apes, eventually given scientific status by Darwin. However, "de-hominization" has also recently acquired some scientific status. The Dutch ethologist Adriaan Kortlandt has formulated a "de-humanization theory" to explain why some chimpanzees used tools and others didn't.

Kortlandt's view is that the chimpanzee is basically a savanna ape that gradually retreated to the forest to flee the expanding human tide. Through observations and clever field experiments, this astute zoologist has shown that those chimpanzees that live near the edge of the forest and roam the cleared areas use various projectiles— rocks and lumps of dirt—to fight their opponents. They even use sticks as clubs or spears to attack leopards, their hereditary enemies, and even to kill small prey, such as guenons (*Cercopithecus sp.*). In the savanna, chimpanzees are occasionally carnivorous.

On the other hand, that does not happen within the forest. For strictly mechanical reasons, weapons are useless in dense vegetation; it's impossible to throw a rock with any hope of hitting the target through a screen of branches, bushes, aerial roots, and vines; it is even impossible to swing a stick. Within their forest retreat, chimpanzees are thus driven to an ever-stricter vegetarian diet, able to catch only minuscule prey (insects, spiders or small soft invertebrates), and at best frogs and small lizards. Chimpanzees, as active predators, had the beginning of tool-making abilities and seemed to be moving towards hominization but, by hiding deep in the forest, they became de-hominized.

Pursuing Kortlandt's reasoning, one might consider the gorilla, a close relative of the chimpanzee living only in the forest and an even stricter vegetarian, as the end-point of that process.

As we already know, a similar shift in habitat is responsible for the incredible transformation of skilled Mousterian hunters and toolmakers into today's brutish pongoid men.

But is this just such a special case? A similar de-hominization seems also to have taken place among Archanthropians. At one time, approximately corresponding in Europe to the beginning of the Mindel glaciation, the Trinil Pithecanthrope ("Java Man") seems to have been ignorant of the use of fire and stone tools, whereas his close contemporary Peking Man left remains of fire pits and stone tools, as did their common North-African ancestor, the Ternifine Man. It is generally thought that Java Man might have lost his technical skills because of his island isolation.

Even among Australopithecines there is some evidence for such de-hominization. *Paranthropus* (*Australopithecus robustus*), living in mid-Pleistocene, did not produce stone tools, while its predecessor in the early Pleistocene, *Australopithecus africanus*, left complex bone-tooth-and-horn (osteo-dento-keratic) artifacts. It is due to his gradually becoming a peaceful vegetarian that *Paranthropus* would have lost his skill at tool making, in contrast to its more agile predecessor, an omnivorous predator.

So, it's in all humanoid lineages, i.e. from a Darwinian perspective, at all stages between the anthropoid apes and modern man, that this de-hominization phenomenon has been observed. It is then a common tendency rather than some exceptional anomaly.

One should also note that this cultural decline is generally accompanied by anatomical transformations best described by the term "bestialization." The forehead recedes even more, the jaws grow bigger, and the enlarged chewing muscles lead to growth of the cranial bony crests where they attach. Their overall appearance may also change: the head sinks into the shoulders and the posture gradually leans forward, tending towards a horizontal, quadrupedal stance. Not only do all those creatures undergoing de-hominization stop behaving like Men, they resemble more and more the image that we have of a Beast.

Such an anatomical de-hominization among Neanderthals has long been recognized. Already in 1906, Thomas Huxley, Darwin's most enthusiastic supporter, had said about the first known Neanderthal remains: "At best they indicate the existence of a human about whose skull we might say had reverted in some way towards an ape-like stage."

As we will see, the label "ape-like" is hardly justifiable in that case. However, it has become fashionable today to speak of a devolution or return to the past when describing the evolution of Neanderthals. The general tendency among zoologists emphasizing genetics is to attribute such degradation to genetic drift, the random accumulation of small unfavorable mutations within a small population.

Could one still imagine an unfortunate incident, a real turning back that would indeed be an exception in evolution, when we discover that it turns out to be a process commonly found among Hominoids? In that case, one should also find a similar reversal in the evolution and specialization of all those vertebrates that have returned to marine life, including reptiles, mammals, and even birds. Even though they certainly returned to the sea, who would think of claiming that ichthyosaurs, dolphins, and penguins became fish in the zoological sense of the word? Travelling from one medium to another, they all followed their own evolutionary path, *progressing* in their adaptation to the conditions of their new life. They did not retrace their steps. At best, we can say that they took a side path, a parallel road.

Similarly, de-hominization, meaning a gradual distancing from those traits that characterize *Homo sapiens*, reflects a normal evolutionary tendency among Hominoids, a tendency from which only Man, in the strict sense, has escaped, or, somewhat more subtly, not succumbed to.

I would call such a widespread tendency the "easy-way-out" solution. It would be most readily compared to the above-mentioned tendency showed by terrestrial vertebrates to return to the ancestral ocean. The conquest of dry land had for all of them been a hard struggle. Once some amphibians had turned into reptiles and had freed themselves from the aquatic medium, it would have seemed that the game was completely over. Nevertheless, at all stages of reptilian and later of mammalian evolution, certain lineages that we would consider lazy or defeatist returned to the sea, a more comfortable, larger environment that's richer in food; in a word, an "easier" place to live. It all seems to happen as if some lineages abandoned a long and exhausting struggle in a new, strange and hostile environment.[64]

In the entire Hominoid family—moderately large primates needing ample space and an abundance and variety of foods such as are found in the savanna—one finds a similar phenomenon: a frequent abandonment of the search for new territories to conquer in favor of a return to the easy life in the ancestral forest.

In order to survive when still diminutive creatures with no other defense than hiding, primates had to be born and develop in deep, impenetrable forests, full of hiding places, and where flight is possible in all three dimensions. Many, although

---

64 Such a phenomenon is readily explained from a genetic point of view, given the small probability of simultaneous development of multiple adaptive devices required by the conquest of a completely different biotope. Evolution always proceeds by trial and error.

not all, became tree-dwellers. The most prudent, the tarsiers and lemurs, sought greater safety in a nocturnal lifestyle. Among the latter, some would not venture in daylight until they found safety in an island poor in predators, Madagascar. On the warm continents, the other primates grew in size and began to feel constricted in the forests and hungry for a broader diet; soon at least two lineages ventured forth in the savanna: the hominoids and the dog-faced baboons. The latter have managed to remain in the open spaces to this day, thanks to their large size, their sharp teeth, and their social organization. There were some defections, however, notably the mandrill, which preferred returning to the forest, in spite of its large teeth, perhaps because of its lack of social organization.

There were also early retreats among Hominoids. The anthropoid apes were most likely the first to give up the conquest of open spaces and the opportunity to hunt the large game living there; among them, the chimpanzee was undoubtedly the last to return to the forest, preferring to stay at the edge of the woods so as to take refuge when needed, or at least rest there peacefully. Among the Australopithecines, Paranthropes apparently followed the same path and perhaps chose a life in the forest as peaceful as that of today's gorillas, while their predatory cousins armed with tools made of bones, teeth, or horns vied with small men (*Homo habilis*?) armed with sharpened stones and were apparently eliminated. I am tempted to believe that Archanthropians, slender and swift as they were, faced the same dilemma: either find refuge in the jungle, preferably in a remote island like Java and abandon all tool-making, or compete against men of the lineage that led to the Sapiens, finally to fall victim to the latter's superior technology and social organization. The extremely late survival of the Broken-Hill Archanthropian might be explained by its prudent retreat into the tropical forest.[65] Similarly, the Pithecanthropes recently discovered at Kow Swamp, Australia, which are only 9,000 to 10,000 years old, might have survived because they reached the Australian continent before the arrival of the first Sapiens and were only joined there later by their implacable rivals. From what we now know about Neanderthals, there can be no further doubt that one day they gave up competing with the Sapiens for their territory and fled the plains, the equivalent of the savanna in the area where they lived. For lack of impenetrable forests of the tropical type in the Palearctic region where they lived, or of suitable islands, they took refuge in the most unwelcoming areas, despised by the Sapiens in their thirst for conquest and invasion, namely the mountainsides, the frozen taiga, the steppes, and arid deserts, and, sometimes some relatively warm high-altitude forests such as in

---

65   I wouldn't be surprised if the persistent rumors concerning the existence of wild hairy men in the forests of the Congo were based on the survival of that form to this day. As the prominent animal trader Charles Cordier mentioned, these creatures are called Kibomba by the Bakano; the Bakondjo, Apamandi, or Abamaanji by the Bakumu; Tchigombe by the Batembo; and Zalazugu by the Warrega.

southeast Asia. Finally, even today within the ranks of the Sapiens species, faced with the expansive tendencies of the conquering, most technically advanced races, some populations are retreating into the tropical forest, shunning material progress, but benefiting from some degree of security and a rather uncomfortable but easy lifestyle, returning to the natural stage of hunting and gathering, and thus suffering a "cultural de-hominization." Examples include the Negroid pygmies of Africa, the Semang Negritos and the Senoi Veddoids of the Malacca peninsula, and the Amerindians of the Amazonian forest.

Most likely, over time, the de-hominization process must have passed from an initial, strictly cultural phase, to an anatomical phase, a morphological difference that became more and more important over time. Today, for example, the Amazonian Amerindians are de-hominized only by their cultural lag. Among African negroid pygmies, the process is beginning to show physical features, in particular a diminution of stature. Among extreme Neanderthals, the process is clear, from head to toe, in the entire body anatomy. Among Archanthropians it is just as advanced, if not further, especially in the development of the brain and the skull, and even more so among Pithecanthropians returned to the wild, in Java, than among the tool-making Peking Man, and also more so among herbivorous *Paranthropus* than in their meat-eating predecessors. Finally, among living anthropoid apes, the physical transformation is total, especially because they have closely adapted to a tree-dwelling life. Thus, the de-hominization is most striking in those races that departed most early from the Sapiens lineage to become separate species.

Since de-hominization has always occurred, at all levels of development of hominoid primates, doesn't that suggest that it is the *normal direction of evolution*? To become less human, one has to have been somewhat human to start with. Might all Hominoids, including the anthropoid apes, perhaps be descendants of some kind of human?

The general de-humanizing trend tends to confirm what the study of embryonic development should have led us to suspect, namely that *Homo sapiens* is anatomically primitive with respect to the other hominid species. The specific lineage to which it belongs must then be very ancient. That far remote ancestor, common to all Hominoids, must have had those traits most characteristics of Man: round head, upright posture, flat non-prehensile foot. And, since evolution is usually accompanied by an increase in size, that ancestor must have been quite short.

We are now back to the "homunculus" theory of human origins, previously mentioned in Chapter 3, where I pointed out the arbitrary nature of the Darwinian scheme according to which Man would descend from Apes.[66]

In this work, which challenges such a solid anthropological belief as the extinction of the Neanderthals, I would have preferred not to also bring in a rather heretical theory of human origins. But that can't be avoided. It should have been expected from the pen of a disciple of Dr. Serge Frechkop. Those who are familiar with his work are aware of my former master's preference for non-ape theories of human origins, including those of Ranke, Kollman and Osborn, and especially Max Westenhöfer's theory of initial bipedalism. For over thirty years I have mulled over these ideas, based originally on the study of embryology and comparative anatomy, and I find that every new discovery in paleontology has confirmed their soundness. I am well aware that my insolence in defending these theories here will bring as many sarcasms, critiques, and even insults as my candid description of the frozen specimen of a contemporary Neanderthal. It would have been so much easier to fit that discovery within the currently accepted anthropological framework. However, Science need not be concerned with Diplomacy: its only duty is to present the facts and to interpret them in the light of all other known facts. So it doesn't seem possible to me to fully understand pongoid men—the logical end-point of the de-hominization of the Neanderthal lineage, long thought to be regressive, retrograde, aberrant, and degenerate—without linking this process to a more general one and to fit it in coherently within the normal course of primate evolution.

While de-hominization is a fundamental tendency of primate evolution, one cannot deny, as you might point out, that there exists also a "humanization" tendency that brought about the emergence of Man proper, *Homo sapiens*. That tendency is also found among other primates. It is a fact that anthropoid apes seem more human-like than monkeys with tails or with dog faces. Being bipedal, Australopithecines seem more human-like than chimpanzees and gorillas and, given the size of their brains, Pithecanthropes are clearly more humanized than Australopithecines and Neanderthals in turn more so than Pithecanthropes. There is no doubt about it. However, real hominization is the result of a set of very specific, sometimes very subtle, modifications.

It's his brain that characterizes Man. However the cerebral evolution that led to the "human phenomenon" was not merely an increase in size of that organ. Three factors were important: size, shape, and fine structure.

A large enough brain is essential for the development of a high level of intelligence, linked to the complexity of the brain: one can't, of course, stuff as many neurons in

---

66   That theory of Darwin's has faced many objections, most of them for the wrong reasons. It was originally rejected because it was thought to be undignified for Man to be a descendant of a Gorilla or an Orangutan. Personally, since I have more sympathy for apes than for people, I would have been delighted if that were the case, which it isn't. Too bad! Our real ancestor was of the cloth that made him the Scourge of the Planet rather than the Lord of the Jungle.

the brain of a marmoset as in that of a man. However, the level of intelligence is not directly proportional to the size of the brain: if that were so, whales and elephants would be much smarter than people.

The shape of the skull influences the development of the different zones of the cerebral cortex, which control sensory and locomotor functions. A brain stretched out in the fore and aft direction, like that of the anthropoid apes, and to a lesser degree that of Pithecanthropes and Neanderthals, favors the development of different zones than the taller and nearly spherical skull of Man. More precisely, that stretched-out shape inhibits the spreading of the frontal lobes, the seat of enhanced intellectual capacity.

Finally, the absolute number of neurons within the cerebral cortex is an important factor in attaining intellectual superiority. The orangutan's cortex, three times smaller than that of Modern Man, holds a billion neurons. A similar but larger and heavier anthropoid, such as perhaps Peking Man, would have had a brain as large as Modern Man, holding perhaps three billion neurons. It wouldn't be as intelligent: the human brain counts more than fourteen billion neurons. Besides their greater abundance, the number of interconnections is the last step in achieving an elevated level of cerebral activity.

The fine-tuning of that complex mechanism was obviously stimulated, perhaps even caused, by a variety of ecological conditions (a favorable climate, open spaces, the abundance of big game, the availability of materials suitable for tool-making, etc.) and behavioral traits (gregariousness, social organization). Without the delicate setting of cortical neurons that eventually enabled the development of articulated speech, Man would never have emerged. The sight of pongoid Neanderthals shows that one may well look like a man anatomically, in zoological terms, without being a man, in philosophical terms.

So there is undoubtedly a subtle and progressive process of hominization among the higher primates, but how can it be reconciled with the even more evident contrary process of de-hominization? Actually, both processes act simultaneously, but in parallel, within different areas of the body and at different levels of organization. On the one hand, there has been a macro-evolutionary de-hominization at work on the general anatomy, clearly evident for example in the skeleton, and on the other a micro-evolutionary hominization process focused on the brain and working at the cellular level, most clearly seen in behavior.

Of course, these opposing processes are not entirely independent of each other. The bipedal posture favors, for mechanical reasons, the development of the brain at the top of the spinal column, but at the same time, the increasing complexity of the cerebral system requires an expansion of the brain, which affects the shape of the occipital region; the shape of the skull must then be balanced with the rest of the body's architecture.

In summary, each type of Hominoid, from anthropoid ape to Man *or vice-versa,* is the result of the level of evolution reached by one or the other of these evolutionary processes, or in other words, the meeting point of anatomical de-hominization with a certain level of cerebral hominization. At every bifurcation of the lineage leading from the original homunculus to modern *Homo sapiens,* de-hominization kicked in when some level of cerebral hominization had been reached, a process that then continued to the extent that anatomical de-hominization allowed it.

Starting from a small round-headed bipedal creature, anatomical evolution led to a fan-like[67] profusion of forms trending more or less towards a morphological stage similar to that of ordinary monkeys, so called Cynomorphs (meaning "dog-like" since they are quadrupeds, with a long muzzle and, usually, a tail). Of all those forms, *Homo sapiens* is the most primitive and the anthropoid apes the most evolved. Between these two extremes, one finds the Australopithecine, Paranthropian and Neanderthal lineages, which have all evolved, at different levels in different parts of their body, towards a more "simian" stage. It would actually be more accurate to speak of a "less-human" stage, since the anatomy of monkeys is characterized by an adaptation to tree dwelling, and one cannot really say that the various lineages mentioned above have evolved in that particular direction.

In comparison with its remote ancestor—the hypothetical homunculus whose remains will most likely eventually be found in Eocene, or perhaps even Paleocene, deposits—modern Man has greatly increased in stature and become more slender. His face has developed only slightly in comparison to his occipital region, adding only a chin. His hand has improved to the point that his thumb can now be opposed to all the other fingers, making him a skilled manipulator. His foot has stretched and become narrower and more arched; its mechanical axis has shifted towards toe No.1, which became a markedly big toe in comparison with the others, which have shrunk. In short, his foot has become specialized for running on flat ground.

The anthropoid apes are the Hominoids that have experienced the greatest modification since the original gnome. In the trees of the forest, they have gradually been transformed into expert trapeze artists. Their arms, used in hanging from branches, lengthened; their thumbs lost opposability and their hands morphed into hooks; their legs atrophied somewhat; their feet became just as prehensile as their hands when their big toe became opposable. Able to proceed along horizontal branches by using their four "hands," they fell back on all fours, without becoming full quadrupeds like the Cynomorph monkeys. The great apes' distinct origin shows itself by the way in which they lean on their extremities when they walk on the ground: they do not walk on the palm of their hands like the cercopithecines or the baboons; they lean on their knuckles. The forward tilt of their torso has affected their balance and the carriage

---

67   The image of a fan is of course only schematic; it is inaccurate since it is the projection on a plane of a much more complex three-dimensional pattern.

of their head, and required the rearrangement of their skull structure; the occipital foramen has migrated towards the back, favoring the forward development of the jaws, while the growth of the brain, wedged between the muzzle and the cervical vertebrae, was stopped. Their forehead became more receding and their prognathism more pronounced. From a nearly circular (actually closer to parabolic) pattern, their dentition has deformed into a U allowing frontal development of the canines, which have become fearsome fangs.

The Australopithecine lineage appears to have been the first after the Pongids to split from the ancestral hominoid main stem, but with more difficulty, having also evolved further towards hominization in the mean time. That is why the dentition of its members is so human-like. Professor G. Vandebroek, of Louvain, a prominent specialist of dental evolution, wrote: "The Australopithecines' dentition is in some respects more human-like than that of Pithecanthropes." On the other hand, they soon acquired a number of anatomical traits converging on those of the anthropoid apes: more or less pronounced prognathism depending on their diet, receding forehead, cranial sagittal crests. The latter were most pronounced in those Paranthropes, which opted for a vegetarian life in the forest, becoming a near pongid.

The Pithecanthropes seem to have been as ancient, or nearly so, as the Australopithecines. Just as with the African Australopithecines, they remained for a long time savanna dwellers, as evidenced by their long and straight legs, very similar to those of Sapiens. Perhaps their rapid specialization into first class runners was achieved at the expense of the development of their brain, already quite large, as befitted descendants of a hominizing lineage. Among them as well, the skull became more strongly receding, particularly so within the Java island refuge, where the absence of serious rivals lessened selection pressure, and where the impoverished genetic pool of an isolated population accelerated the evolution towards a more beastly form.

Neanderthals, the object of our specific interest, appeared much later, during the Mindel-Riss interglacial, at a time when the Australopithecine hordes had disappeared from Africa and the Archeanthropians had spread out and differentiated throughout the warmer regions of the Old World and were already in decline. The paleanthropians seem to have begun to split away from the Sapiens lineage when the latter's feet and hands were still at a relatively primitive stage. Indeed, among some of them, the thumb is not yet completely opposable, and the axis of the foot is still nearly median (as in the ancestral fin) and passes between the second and third toes. But certainly the pre-Sapiens that gave rise to the Neanderthals had already reached a high level of cerebral development when the split occurred since the volume of the brain is on the average greater among Paleanthropians than in Modern Man. We also note that with respect to the development of their nose, which is sometimes described as "ultra-human," Neanderthals have gone beyond *Homo sapiens*. In contrast, they are much more simian looking because of their receding forehead, bowlegs, and bent-over stance.

Actually, the Neanderthals' evolution is only apparently simian. They experienced an adaptive evolution in a direction quite different from that of the apes. From the very beginning of the Würmian, they transformed so as to be able to withstand extreme cold and to adapt to a rock-climbing life, while the anthropoid apes took refuge in warm regions and adapted to a life of tree climbing. The flat, pushed-forward face of Neanderthals (oncognathism) is quite different from the prominent jaws of the apes (prognathism). Their gripping foot has nothing in common with the apes' foot with its prehensile, grabbing big toe. It is a mere coincidence that life in the mountains and life in the trees are both compatible with a more forward-leaning posture than life as a runner in the plains. De-hominization occurs just as well by retreating to inaccessible mountains as by hiding in the forest.

Anyway, it would be just as false to say that Neanderthals are ascending to the apes as they are descending from them. If I had to define most appropriately the paleanthropian evolutionary tendency, I would say that *Neanderthals are to Sapiens what a bulldog is to a ordinary dog.*

Everyone agrees that Man is a domesticated animal since it has been artificially selected by Man; we are a self-domesticated species. Now, a variety of mutations, always the same, appear among domesticated mammals: gigantism, dwarfism, albinism, melanism, piebald coloring, hairless skin, etc. One of the most peculiar is a mastiff-face mutation, which is characterized by a flattening of the head and twisted limbs. It is not only commonly found in dogs (mastiffs, boxers, Pekinese, etc.) but also among cattle, swine, sheep, goats, and even cats. But, aren't these the same kind of deformations observed among pongoid men? I believe that the specialization of the Neanderthals was triggered and facilitated by the appearance and the intensification of a mastiff-face gene among some individuals of the Sapiens lineage who eventually split from it under the selective impact of glaciations. It so happens that this particular mutation transforms the human body so as to render it more capable of enduring extreme cold, as well as enabling its progress in steep terrain. Just as the melanotic mutation was favorable in sunny places, or dwarfism within the tropical forest, which respectively led to black-skinned races and pygmies, the mastiff-face mutation was a favorable adaptation to cold and mountainous areas, which gave rise first to a Neanderthal race and eventually the Neanderthal species.

Let's now consider the overall evolution of Hominoids. We note that an apparent de-hominization occurs in all lineages, except that of Sapiens, after a specialization. Ernst Mayr astutely wrote in 1962: "Man, one might say, is specialized in de-specialization." Perhaps it might be even wiser to say, with Lorenz (1959): "Man has specialized in non-specialization." That is what has allowed him to successfully occupy an incredible variety of terrestrial ecological niches (except, apart from exceptional circumstances, those occupied today by the relic Neanderthals and pongoids). All others have specialized. Anthropoid apes have become forest trapeze artists, Australopithecines savanna hunters, Paranthropes powerful vegetarian brutes, and Neanderthals expert alpinists. It is not clear in which direction Pithecanthropians became specialized, but their basic progress might have been the acquisition of greater mobility, of superior maneuverability.

Any specialization, of course, takes place at the expense of some other abilities. Pongids (the Orangutans) lost their bipedal stance after evolving in three dimensions; the Australopithecines were too busy strengthening their mandibles to solve dietary problems and lost their spherical head shape; the Archanthropians followed in their footsteps by perfecting their locomotive powers: Paleanthropians lost in their mountains their vertical posture and communal life. *Homo sapiens* is the only one left who benefited from all the advantages favoring the development and the improvement of the brain. The only one not to suffer de-hominization, he is also the only one who was capable of acquiring an articulated language, with all its consequences.

It is the accumulation of cultural progress made possible by articulated speech that made the lightning acceleration of technical progress possible. And it is undoubtedly the scale of its adaptation, exploitation, and even destruction of its environment that best characterizes the human species relative to the rest of the animal world, happy with its environment as it is.

Of course, the other living creatures, particularly the most closely related rivals, are also part of the environment. One of the most striking characteristics of *Homo sapiens* is its drive to destroy or subjugate, and even, beyond all its basic needs, to persecute all its rivals. That concern about eliminating brothers and cousins occupying potentially available ecological niches is actually a primordial factor of hominid evolution. De-hominization is the result, in most lineages, of a flight to inhospitable areas, such as mountains and forests, before the rising tide of the fearsome and merciless representatives of the lineage leading to today's Sapiens.

Such an idea was untenable as long as one believed in a direct succession of various types of Hominoids. However, paleontological discoveries have finally clearly demonstrated the lengthy overlap of different lineages.

At the beginning of the early Pleistocene, creatures as clearly differentiated as *Homo habilis*, the presumed ancestor of the Sapiens lineage, *Australopithecus africanus,* still very human-like in its predatory behavior, and good old *Zinjanthropus,* an early de-hominized vegetarian brute, coexisted in South Africa (and perhaps also in Ethiopia).

Later, somewhere between the early and mid-Pleistocene, the Kromdraai and Swartkrans Paranthropes, as well as the Telanthrope (a *Homo habilis*), all lived in South Africa, while Archanthropians were found in Java (Djetis) man and China (Lantian).

At the beginning of the mid-Pleistocene, there were many Archanthropians in Africa, such as the Atlanthropes of Ternifine in Morocco, the Olduvai Gorge chelleans men in Kenya; Archanthropians were also found throughout the warmer parts of Asia, from the Sinanthropes of Chou-kou-tien in China, to the Trinil Pithecanthropes of Java. In Europe, early pre-Sapiens were present in Vertessözllös, in Hungary.

In the heart of the mid-Pleistocene, during the Archanthropian expansionist era, there were already pre-Sapiens in Africa (Kanjera) as well as in Europe (Swanscombe).

And at the end of the mid-Pleistocene, while the pre-Neanderthals were spreading through all the cold and temperate regions of Eurasia and North-Africa, other than a few pre-Sapiens (Fontéchavade), there were still Archanthropians in the tropics, in Ngandong on Java, as well as in Dire-Dava and Omo in Ethiopia.

Finally, at the end of the late Pleistocene, while the conquering Sapiens had invaded all parts of the Old World, chasing the Neanderthals into colder regions, Archanthropians still survived in Zambia (Broken Hill) and Australia (Kow Swamp), and even a few Australopithecines, such as Coppens' Chadanthrope in the heart of torrid Africa only 10,000 years ago.[68]

It is obvious that all these species must have clashed and fought during their expansion, at least whenever they competed for the same habitat.

Archanthropians and Paleanthropians probably didn't interact, since the former evolved in warm southern regions, while the latter developed in cold northern climes. Each type spread out over ever-wider areas, like two allopatric races adapted to different environments. The only places were they would have been likely to meet were in southeast China and North Africa, but they don't seem to have been there at the same time. The Neanderthals, who came later, couldn't have shown towards the Pithecanthropians the same spirit of competitive intolerance with which the Sapiens treated them. Not having known each other, they could not hate each other.

Things were quite different for those on the way to becoming Sapiens. Precisely because of their lack of specialization and their ability to adapt to a wide variety of environments, they coveted them all and did everything they could to displace previous occupants. Other Hominoids, more restricted in their choices, did not always see pre-Sapiens as rivals, although the latter saw all others as competitors.

The predatory Australopithecines seemed to have disappeared before the Ice Ages, while the inoffensive vegetarian Paranthropes survived at least until the end of the early Pleistocene. Perhaps they were the first victims of *Homo habilis*, who was already wielding stone weapons.

We know very little about the possible interactions between Sapiens and Archanthropians, although some authors have suggested that the Sinanthropes might have fallen prey to Sapiens, who were already living in the same region at the time, a plausible suggestion. On the basis of what we know today about the merciless

---

68   Cf. Servant, M., P. Ergenzinger and Y. Coppens, 1969. C.R.S. Soc. Geol. Fr., Paris.

persecution of the Neanderthals by Sapiens, there is no reason to believe that Sinanthropes, Atlanthropes and Pithecanthropes would have been spared a similar fate.

It is thus most likely that the halt of Archanthropians' expansion across Africa and Southeast Asia, their subsequent pushback towards Indonesia and Australia, and the final confinement of their habitat were caused by the extension of the pre-Sapiens.

Finally, we are already aware of the fate that the Sapiens dealt the Neanderthals over a series of assaults during interglacial periods.

So it appears than beyond a legitimate wish for survival in occupying and defending a territory, an irrepressible thirst for butchering rivals and bystanders would be a dominant psychological characteristic of the *Homo sapiens* lineage, at least as much as technical ability. I would describe this tendency, found in no other animal species, by the zoological term "hypertelic aggressiveness," an aggressiveness that goes beyond its goals. Hyperthelia, which seems to result from inertia of an evolutionary process, is commonly seen in the animal kingdom. Cases of hypertelia include the extraordinary development of the Irish stag's antlers, or the rolling up of the mammoth's tusks, neither of which could perform their original functions and ended up being a nuisance to their owners, contributing to their extinction.

Within the Hominoid main stem, the population that has carried the strongest contribution of that excessive aggressiveness in its genetic baggage has eliminated, or at least tried to get rid, first its closest relatives, then its neighbors, and may finally some day eliminate itself.

Sticking to the most recent and best-documented facts, the Sapiens first practically eliminated the Würmian Neanderthals. Then, within their own species, the Caucasians, or Whites, who spread over two thirds of Eurasia, pushed back the Bushmen and Capoids towards the east and south of Africa and the Congoids, or Blacks, towards the western part of the continent. Meanwhile, the Mongoloid, or Yellow race, spreading over the remaining third of Eurasia, invaded the Americas and drove the Australoids to Indonesia and later to New Guinea and Australia. Within historical times, there were some attempts by the Mongoloids to re-conquer part of Eurasia, but over the past few centuries the wave of colonial expansion of the technically superior Caucasians has submerged the whole planet, bringing the near extermination of whole populations, such as the Bushmen, the Australian aboriginals, the Maoris, and the North American Indians. Even today this orthogenetic process continues, relentlessly. South American Caucasians, even of mixed race, systematically massacre those Indians who remained hunter-gatherers. Even within the great Caucasian race, there have been efforts to eliminate some ethnic groups, such as the Jews and the Gypsies. In 1960, L.F. Richardson, an Englishman, calculated that in 126 years, between 1820 and 1945, 59 million humans had died in wars, revolutions, and organized massacres or from assassinations or bloody riots.

Homo sapiens deserves the nickname *Homo interfector* just as much as that of *Homo faber,* Man the Killer, as much as Man the Toolmaker. It is essential to grasp that truth if one is to understand the history and evolution of Hominoids, and indeed of the living world as a whole. The study of the pongoid Neanderthals adds a new chapter to this issue, which was first raised by Lorenz.

Man's remote ancestors could, of course, be excused. They were extraordinary helpless creatures. Small and frail. Without a shell, one couldn't be more naked. Without claws or fangs, equally incapable of attacking or defending themselves. Even incapable of running away: they couldn't gallop away, or fly up in the air, or quickly climb up trees. Condemned to hiding in holes or in bushes. How were they to survive such handicaps?

To overcome them, they had to be very clever. There were also benefits to be had in grouping together and organizing; union brings strength. But the ideal remedy came with their manual dexterity; for lack of natural weapons, they had to make artificial ones. And, mainly, in order to take full advantage of those skills, they had to be aggressive, greedy, and dominating. *Only those most generously endowed with intelligence, social sense, dexterity, and aggressiveness had any chance of surviving.* Those were indeed the virtues that, through the action of natural selection, were developed and emphasized. And, as often happens with evolutionary processes, the mechanism got carried away with its momentum, with well known consequences: intelligence wasted and powerless to solve the problems it created, an insane level of overpopulation defeating all efforts at social organization, and a self-devouring macho attitude leading to gratuitous violence and auto-destructive rage.

Of all the larger mammals *Homo sapiens,* omnipresent and horribly prolific, is nearly the only survivor among the last herds of free animals, massacred without mercy or reason, on a planet that has been pillaged, polluted, and deforested. We should be humble! The sum of human achievements is nothing to rejoice or be proud about.

In conclusion, if I may express a wish, it is to see humans at last become civilized by protecting the pongoid Neanderthals they have persecuted for millennia. I hope that there will be created for that purpose, numerous protected reserves where our unfortunate brothers will be left to live in peace. They are perhaps the image of what we should never have stopped being, Man before the original sin, the sin of suicidal technical skill and boundless aggression.

Abel is not dead. He recovered from his wounds. Today, we can choose not to finish him off, not to assassinate him once more. Will we not be touched by the eye, kind and fearful, with which he looks at us from a tomb we have already dug? That would allow us at last to escape from the guilt of Cain, the killer, to break away from the curse that haunts our species.   — Bernard Heuvelmans
Finca "El Salto" (Escuintla), Guatemala, April 1969
Ile du Levant, France, August 1973

Afterword

After The Thaw: The Post-Heuvelmans Iceman
by Loren Coleman, 2016

It's been almost 50 years since Bernard Heuvelmans and Ivan T. Sanderson first laid eyes on the so-called Minnesota Iceman, and much has changed over the decades in our portrait of this alleged hairy hominid carcass that Heuvelmans baptized *Homo pongoides*. But first, let me recount how my life first intersected with the now-famous Iceman.

My Personal Moments With the "Body"

I became involved in cryptozoology research in 1960 and soon struck up a correspondence with Bernard Heuvelmans and Ivan T. Sanderson. Sanderson introduced me to fellow cryptozoologist Mark A. Hall, only a year older than myself. Mark was born in 1946 and raised in Minnesota. I grew up in Illinois.

Loren Coleman (left) and Mark Hall in 2003,
34 years after they investigated the Minnesota Iceman traveling exhibit.

After having spent three days studying the Minnesota Iceman in its frozen enclosure in December of 1968, Heuvelmans and Sanderson alerted us to be on the lookout for exhibitions of the supposed Iceman traveling in our area.

It soon became clear that Hansen had switched the original body for a model in the spring of the following year. Writing in 1999, Mark Hall observed that in April 1969, "Hansen canceled his commitments to display the Iceman and arranged new dates and places to exhibit his model. The first date and place was the Midway Shopping Center in St. Paul, Minnesota, on 5 May 1969. I was there to see the exhibit. I met Frank Hansen and his wife. I requested permission to photograph and did so." Then, Hall notes, "Hansen took his model on a tour of Midwestern states."

On Saturday, August 8, 1969, I saw what was then being called the "Minnesota Iceman" at the Illinois State Fair, in the state capital of Springfield. (The 1969 event, an 11-day fair, held the highest attendance mark for any Illinois State Fair, 1,155,304 people, until that record was broken in 2013, with 1.2 million, according to *AgriNews*.) At the exhibit, I spoke to Frank Hansen briefly and went through the exhibit twice.

I took photographs and sent them to Sanderson, Heuvelmans, and Hall. The four of us worked closely together to determine the differences between the hominid-like body that Sanderson and Heuvelmans had seen and photographed in December 1968, and the obvious fake that Hall and I saw a few months later in 1969. The toothy model that is often shown as the "Real Iceman" is one of the photos taken of the acknowledged fake being shown in 1969.

"When Ivan Sanderson saw those pictures," wrote Hall, "he was able to say that the model was quite different from the Iceman. When a copy of Heuvelmans' paper in the Belgian journal became available to me in July [1969] I could see what he meant. It was similar but it was not the same thing."

Hall is clear about what has happened to the Minnesota Iceman: "Except for the photographs taken by Heuvelmans in 1968, all photos that are supposed to show the 'Iceman' date from April 1969 and after. They do not show the 'Iceman' but instead show a model that Hansen displayed after April 1969."

Sanderson and Heuvelmans discussed 15 specific differences they noticed between the original Minnesota Iceman and the model. They include known details for the original body, such as vegetable material apparent in the teeth, parasites on the skin, and other items mentioned by Heuvelmans in his book, which were not there on the model.

My life would again cross paths with the alleged "Minnesota Iceman" years later, in 2015, but more of that in a moment.

The Discovery, Media Storm, and the Switch
Ivan T. Sanderson, naturalist, zoologist, cryptozoologist, born in Edinburgh, Scotland, in 1911, had moved to New York City at the end of World War II but remained a

British citizen. Sanderson was a major media personality. He hosted the world's first color television program, and was the first person to bring animals on talk shows; Sanderson was the "Animal Man" for *The Garry Moore Show* and then *The Today Show*. His TV appearances and early cryptozoology books made Sanderson the go-to person regarding any new "cryptozoology" discovery. It was in this context that Heuvelmans regarded Sanderson as his cryptozoological inspiration and mentor. And so it was that Sanderson came to host the Belgian-French zoologist at his home in New Jersey before they departed together to view the Minnesota Iceman.

Sanderson promised Frank Hansen that he would keep the Minnesota Iceman "discovery" secret. He only vaguely and indirectly mentioned the body on *The Tonight Show* with Jack Jones, a Johnny Carson guest-host, on Christmas Eve, December 24, 1968.

Then on January 18, 1969, Sanderson informed the Federal Bureau of Investigation of the matter, and Heuvelmans asked primatologist John Napier to pressure the Smithsonian and the FBI to obtain the carcass.

Sanderson felt he had been on the verge of getting the "damned thing out of the ice" when "that bloody fool Heuvelmans bulldozed the Belgian Academy to rush his version [of the discovery] into print," according to a letter he wrote to Tom Hall of the *Chicago Tribune Sunday Magazine* in 1972. The release of Heuvelmans' science report on March 10, 1969, set off a flurry of media interest in the Iceman worldwide. Sanderson's own account would not appear until May 1969, in the popular men's magazine *Argosy*. Sanderson's formal paper on the unknown hominid carcass appeared later in the year, in the Italian science journal *Genus*.

As news of the Iceman hit the media, S. Dillion Ripley, the secretary of the Smithsonian, contacted Hansen on March 13, 1969. That's when Hansen and the alleged mystery millionaire owner of the original Minnesota Iceman got cold feet. Hansen replied to Secretary Ripley on March 20, 1969: "We have framed an illusion show for 1969 which, in many respects, resembles the specimen photographed by Dr. Heuvelmans while a guest at our ranch. I cannot say with any certainty if the original exhibit will ever be presented again as a public attraction." We do not know what became of that original specimen.

Heuvelmans had hoped that his scientific paper would have pressured Hansen into releasing the original specimen, but Sanderson thought his friend's rush-to-publish would have just the opposite effect. From that point on, notes British geneticist Bryan Sykes "relations between Heuvelmans and Sanderson were never repaired. Heuvelmans saw Sanderson's reputation for sensation as an obstacle to the scientific respectability that he craved, and Sanderson blamed Heuvelmans' publication for Hansen's disposal of the original Iceman. Sanderson continued to blame Heuvelmans for the rest of his life." Sykes feels "the betrayal and the despair" of the Minnesota Iceman affair "destroyed Heuvelmans."

The Secret Millionaire and the Dummy

Mark Hall spent half a year with Sanderson but never picked up hints from him who the specimen's "owner" might be. Then in 2015, Hall, following clues left by Hansen, felt the movie director Robert Wise (1914-2005) might be the secretive millionaire owner of the Iceman. In recent years, Hall researched data and other items about Wise that makes sense in terms of the chronology of the Iceman's story.

As it turns out, Robert Wise was in the right place at the right time. But in this scenario the Iceman came to the USA, not via Vietnam in a body bag, as in Heuvelmans' theory, but via Hong Kong in a filmmaker's luggage, Hall thinks.

The key is the Robert Wise movie, *The Sand Pebbles*, which was released at the end of 1966, starring Steve McQueen, Richard Attenborough, Richard Crenna, and Candice Bergen. Where the motion picture was filmed and the fact that the crew had to bring back tons of equipment provided a cover for bringing back the Iceman. *The Sand Pebbles* was filmed in Taiwan and Hong Kong, beginning on November 22, 1965. It was scheduled to take about nine weeks, but ended up taking seven months.

My own research indicates that Wise liked curios, so he probably sneaked this "ultimate souvenir" back into the US with his filmmakers' equipment. The creation of the model occurred almost immediately upon Wise's return from Hong Kong with the specimen. He hired a Hollywood makeup artist—from the Bud Westmore/John Chambers/Howard Ball circle—to make one of the primary copies. Famous for creating the creature in *Creature from the Black Lagoon*, Westmore worked for Wise

as a makeup artist from *The Flower Drum Song* (1961) through *The Andromeda Strain* (1971). Heuvelmans noted that Hansen and Westmore discussed the Iceman in January 1967. Hansen was viewed as a reliable exhibitor of antique tractors on the circuit. Hansen was never a "carny," as debunkers would have it, but a "showman." Wise probably found that appealing when he decided to "test market" the Iceman on the Midwestern public at state fairs for two years.

In the spring of 1969,

writes Sykes, Frank Hansen "was about to go on tour with a dummy version of the original Iceman, billing it as a deliberate hoax. And indeed he did put something on show in a shopping centre in St. Paul, Minnesota in May 1969 and later in Grand Rapids, Michigan. It was almost certainly not the same corpse that Heuvelmans and Sanderson had examined back at the Hansen ranch in December of the previous year."

Hansen's new "fabricated illusion," as he called it in April 1969, was being shown all over the upper Midwest and into Canada. Once he was stopped at the Canadian/North Dakota border, trying to get home, in July 1969. To the authorities, "Hansen was quite candid about replacing the original with a model," wrote Hall, and Hansen was immediately allowed back in the country.

Things were getting crazy. Hall observed, "The claims to have made models of this thing mushroomed in 1969 to the point where Sanderson wrote in 1970: 'We received five reports on the construction of such "specimens" by model-makers; most notably by the Hollywood, old-time professional John Chambers.'"

Stepping back from these claims, Hall said, "More than one model may well have been made based upon the original Iceman. Hansen has said that he [the mystery millionaire owner] did commission two of them. One of them was discarded as inadequate."

C. Eugene Emery, writing about the John Chambers' associate Howard Ball, said that Ball definitely made a model. Hansen admitted to Emery that Ball had made a copy of the original Iceman, but insisted that it "was discarded."

As time went on, showings of the model appeared less in state fairs, and more and more in shopping malls. Many baby boomers today have shared memories of seeing the "original Minnesota Iceman" on Facebook and in blogs. Most are actually nostalgic remembrances of the later displays in their local malls, after the explosion of publicity in the spring of 1969 through 1972. Those who wished to make a point—pro or con about the Minnesota Iceman—by saying they had seen the Minnesota Iceman, remind me of all the people who have emotionally shared that they were at Woodstock in 1969. "The widespread joke [is] that millions of people now claim to have attended Woodstock, even though the actual crowd numbered about 400,000," wrote writer Paul Lukas in *Fortune*.

The story of the Minnesota Iceman got even more muddled in July 1970, when *Saga Magazine* published "I Shot the Ape-Man Creature of Whiteface" by Frank Hansen. In it, he claimed the Iceman was a Bigfoot shot in Wisconsin. The showman was trying to drum up business.

But the attempt for media attention failed. "The new exhibit did not do well," writes Hall. "Despite the publicity for the mystery, people wanted to see the real thing, not what was openly declared to be a fabrication. Within two years Hansen was farming his exhibit out to car dealers and the like, unattended with only a tape recording and signs to tell the viewers about it."

For years, every time anyone would see the Minnesota Iceman in their local mall, one of my correspondents would send me an article or a note about it. Then in 1994, I was contacted by Cosgrove-Meurer Productions as I had appeared on their NBC program *Unsolved Mysteries* a few times. They wanted to do a segment on the "Minnesota Iceman" and asked for my assistance. Since Hall had been the director of Sanderson's organization, the Society for the Investigation of the Unexplained, after Sanderson's death, and since Hall had been involved in tracking the Minnesota Iceman throughout the Midwest, I recommended the producers get in touch with Mark. They also contacted Terry Cullen for an interview. Cullen had seen the Iceman through clear ice and was the first to tell Sanderson about it.

Sanderson had died at the age of 62 on February 19, 1973, in Blairstown, New Jersey, of cancer, calling for his two golden retrievers right before he passed away. Hansen and Heuvelmans refused to be interviewed or involved with the program. The *Unsolved Mysteries* segment on the Minnesota Iceman was broadcast on September 25, 1994.

The late 1990s and early 2000s would see occasional news reports about the Iceman, mostly from skeptics about the "body" as an elaborate carnival "gaff," a fake claimed to be created by various artists or hoaxers who never were able to produce any photographs of their craft-in-progress of the unforgettable carcass of the original Minnesota Iceman. Claims that Sanderson ever had the body in New Jersey were likewise groundless.

In 1999, Heuvelmans donated more than 50,000 documents, photographs, and specimens to the Museum of Zoology in Lausanne, Switzerland. All of the photographs and drawings he made of his 1968 examination of the Minnesota Iceman are stored there, but few have been able to visit those archives.

On August 22, 2001, around noon, Bernard Heuvelmans, 84, passed away without suffering, in his bed at his home in Le Vesinet, France, with his faithful dog nearby.

Then Frank Hansen died at the age of 80, on March 23, 2003, in an extended healthcare facility in Winona, Minnesota, according to Todd Prescott and Mike Quast.

An era had ended.

The Comeback

Less than a decade later, the long-missing replica of the Minnesota Iceman made a surprising reappearance. Early in 2012, a friendly back-channel communications from Adam Peck, an employee of the person who then owned the model "Minnesota Iceman," informed me that I could buy the exhibit. By October 2012, I was in active negotiations, and received a verbal agreement, to allow the International Cryptozoology Museum to purchase the display on a payment plan.

The new owner was a Minnesota nurseryman, Bill Zywiec, who ran Zywiec's Landscape and Garden Center in Cottage Grove, Minnesota. Zywiec had purchased the Iceman from Hansen's son, and was keeping it in the rear of an outside shed, unfrozen and relatively unprotected. Mold had blackened the model; the Minnesota Iceman was not doing well. It was occasionally being shown during Halloween hayrides. Locals didn't know the historical value of their Iceman.

Then in a surprise move, the owner listed the Minnesota Iceman on eBay early in 2013 for $20,000. I tried but could not come up with the money in short order. At 8:14 PM, on February 20, 2013, the auction ended, with an accepted final bid of $19,000. The winning buyer was the sideshow-oriented Museum of the Weird in Austin, Texas.

The Minnesota Iceman soon became the focus of the A&E reality television program, *Shipping Wars*, in June 2013, when they broadcast the transport of the Iceman from Minnesota to Texas. Steve Busti, the owner of the Museum of the Weird, went along on the trip and discussed the Minnesota Iceman while remaining purposefully vague and a bit tongue-in-cheek about it being a real body.

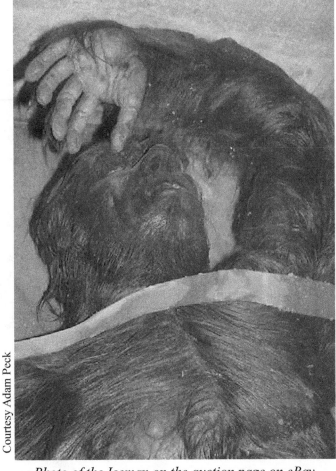

Courtesy Adam Peck

Photo of the Iceman on the auction page on eBay.

The replica of the "Minnesota Iceman" had its grand opening at the Museum of the Weird on July 13, 2013. For an extra charge, visitors could view, in a separate room, the model under ice that's still known as the "Minnesota Iceman."

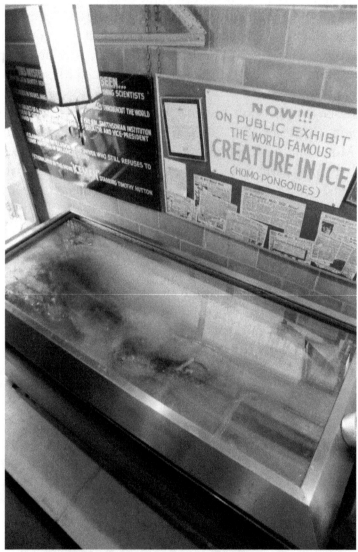

© Museum of the Weird

Steve Busti is a friend, and through a cooperative agreement in which we only had to pay travel expenses, the Museum of the Weird loaned the Minnesota Iceman to the nonprofit International Cryptozoology Museum (ICM), which I founded and direct, in Maine. The ICM hosted this "Iceman," minus the enclosure and the ice, from August 2015 through the first week of January 2016. For the price of an admission to the ICM, visitors saw the Iceman model completely free of ice for the first time.

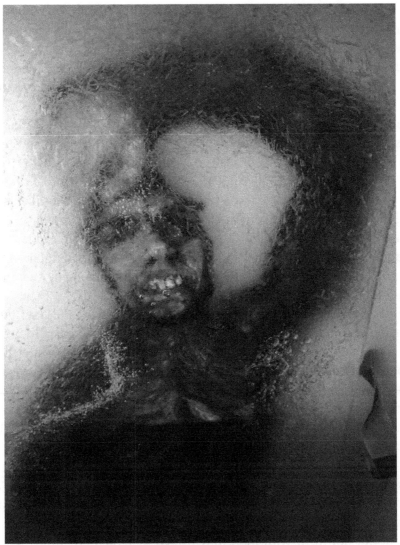

© Museum of the Weird

The model of the Minnesota Iceman is now back in Austin, Texas, and is currently on exhibit at the Museum of the Weird.

By a strange coincidence, in January 2016, I met Terry Cullen for the first time at the first International Cryptozoology Conference in St. Augustine, Florida. Cullen confirmed to me that he too felt the real body was switched for a model during the spring of 1969. He subsequently told me that, contrary to Heuvelmans' statement, he denies having seen the back of the specimen's head, saying only that "the head *seemed* to have a saggital crest when viewed through the relatively clear, thin ice that covered the 'creature'" in his early viewings. So Heuvelmans was incorrect in thinking that "Cullen was in cahoots with Hansen."

The 1974 Neanderthal vs the 2016 Neanderthal

Heuvelmans' book on the Minnesota Iceman was written more than four decades ago. At the time, anthropology was struggling with its own internal battles regarding biases, racism, slanted evidence, quick judgments, and narrow-mindedness. It is in this atmosphere that Heuvelmans attempted to classify what the Minnesota Iceman might be.

Heuvelmans was enlightened, certainly, for he was then one of the new wave of scientists who were firmly rejecting the old, brutish, "cave man" approach to conceptualizing Neanderthals.

In the 1960s, the sloping, hairy, apelike Neanderthals of the Victorian era were being replaced by what must have seemed to be almost hippie-like Neanderthal "flower children." In *Shanidar: The First Flower People*, archaeologist Ralph Solecki wrote: "With the finding of flowers in association with Neanderthals, we are brought suddenly to the realization that the universality of mankind and the love of beauty go beyond the boundary of our own species. No longer can we deny the early men the full range of human feelings and experience."

Heuvelmans clearly rejected the apelike Neanderthal of the day, but still had room in his cosmos for a rather hairy hominid, *Homo pongoides*. Heuvelmans was insightful enough to realize that the picture of the "primitive" Neanderthal was fast growing outdated, and his thinking about matching the Minnesota Iceman with the early Neanderthals was not too encumbered by those old reconstructions.

In a remarkable look at the evolution of how we modern humans view Neanderthals, Lee Rimmer (2016) at *Ancestry* wrote, "Over the last 100 years, reconstructions of their appearance have slowly become 'humanized' with each new revelation about their culture and physiology, culminating in the stunning discovery in 2010 that up to 4% of the genome of all modern humans of European and Asian origin carry Neanderthal DNA, as a result of interbreeding between the two species."

The Victorians thought the Neanderthals' "oversized brow suggested an ape-like ancestor that did not fit in with the biblical idea of God's creation," said Rimmer. One of the early scientific names under consideration for Neanderthals at the time provides an insight into their thinking; *Homo stupidus* was the Latin name suggested by Ernst Haeckel in 1866.

One Neanderthal specimen is, for historical reasons, more important than all the rest. "This is the so-called 'old man' from La Chapelle-aux-Saints in southwest France, found in 1908," wrote Christopher Stringer and Clive Gamble in 1993. "We learn from his history that Neanderthals are what we make them." This Neanderthal is so well known it framed and influenced the paintings and drawings of Neanderthals for decades. Then, years later, it was discovered that individual was severely arthritic.

The movement away from this early concept of Neanderthals took place just before the appearance of the Minnesota Iceman. In 1962, anthropologist Carleton Coon wrote, "The problem can be brought within sight of solution only after we have disposed of the popular image of Neanderthal man, that he was a squat, slouching, low-browed, stupid, and vicious brute, wooing his women by clubbing them over the head and eating his deceased parents."

But one of the major characteristics still tied to Neanderthals at the time Heuvelmans was contemplating the Iceman was that they were hirsute.Analyses in the 1980s finally led to a shift in that view. Even the fictional Neanderthal peoples of Jean M. Auel's series, starting with *The Clan of the Cave Bear*, began portraying Neanderthals as much less hairy than previously. Cultural diffusion of the more modern looking Neanderthal was beginning.

"The extensive body hair and mane felt necessary by earlier restorers has vanished," and now there is an effort "to emphasize the tough, rugged, mountaineering-type body form," observed anthropologist Myra Shackley in 1983.

Twenty years later, in our new century, we began to view the Neanderthals as "neighbors." Typical is a documentary which aired in 2000 that showed Neanderthal as more sophisticated than popularly believed, not covered in thick hair, and wearing clothing made of animals skins. Later in that decade Neanderthal reconstructions finding their way into museums were characterized as "thoughtful," "characterful," and "human." As Rimmer's *Ancestry* article concludes, "You could probably find the same range of phenotypes amongst modern humans in any average town today."

Times do change, and it is a lesson in caution. As noted by Marta Fiacconi and Chris Hunt in 2015, the Neanderthal flower children apparently never existed either. The original buildup of pollen in Neanderthal burials, which was interpreted as flowers left as offerings, instead is today attributed to the action of bees. Simply then, we have to be careful, one way or other, that the political correctness of our time is not being reflected in the evolution of how fossil hominids are being elucidated in their appearances.

Heuvelmans did not have the hindsight of all the new evidence we now have about the Neanderthals. Stocky and strong, in essence Neanderthals are understood today as having been hairless, nearly 6 ft tall hominids, who became extinct 30,000 years ago. One remarkable breakthrough regarding Neanderthal appearance, revealed in recent genome studies, has them being apparently red-haired, freckled, and fair-skinned—nothing at all like the Minnesota Iceman. The genetics of Neanderthals' red hair are associated with the melanocortin-1 receptor (MC1R), which is found on chromosome 16. With clothes on, Neanderthals would not have looked out-of-place at your local shopping center.

The major problem with labeling the Minnesota Iceman or *Homo pongoides* a Neanderthal is that Neanderthals are no longer thought to have been covered in hair. But then anthropologists and paleoanthropologists today could be wrong about that too. After all, while we have discovered some whole, fully haired contemporaries of the Neanderthals—wooly mammoths and wooly rhinoceroses—we have not found —at least not yet—a truly well-preserved, hairy or hairless body of a Neanderthal.

But then again maybe the Minnesota Iceman was not a Neanderthal.

Neanderthal or... What?

Heuvelmans' careful study of the Minnesota Iceman led him to conclude that the carcass was an example of a relict Neanderthal. Heuvelmans theorized that the body actually came from Vietnam. He found good comparisons to the Iceman in the hairy hominid reports known as the Kaptar in the Caucasus Mountains and the Ksy-gihik in Central Asia.

Helmut Loofs-Wissowa, an anthropologist at Australian National University, thinks Heuvelmans was correct. He points to the Neanderthals, represented by the Minnesota Iceman, as the source of surviving relict populations of wild men, the Nguol Rung, he has studied in Vietnam. Loofs-Wissowa's creative work links the semi-erect penis of *Homo pongoides* with what he sees in Paleolithic cave art and other evidence he connects to Neanderthals.

Jordi Magraner also thought the Iceman might have been a Neanderthal. Magraner was a Spanish zoologist who did fieldwork for 12 years in North Pakistan and Afghanistan, throughout the 1990s-early 2000s, in search of the local wild man, the Barmanu. Before he was murdered in Pakistan on August 2, 2002, he collected more than 50 first-hand sighting accounts, and all eyewitnesses recognized the reconstruction of Heuvelmans' *Homo pongoides*. They picked out *Homo pongoides* as their match to the Barmanu from Magraner's ID-kit of the drawings of apes, fossil men, aboriginals, monkeys, and the Minnesota Iceman. Of course, this only proves the Barmanu looks like *Homo pongoides*, not that it was a Neanderthal.

But others close to the original investigation think there might be another answer.

The views of Ivan T. Sanderson, who observed the Iceman in the freezer with Heuvelmans, quickly evolved away from Heuvelmans' theory. Sanderson disagreed with Heuvelmans, thinking that the Minnesota Iceman "most certainly should not be assigned to the Neanderthal race or complex," as he wrote in the Italian journal *Genus*. Sanderson thought it was *Homo erectus*, according to Joshua Blu Buhs, who reviewed Sanderson's papers.

Mark Hall pondered the identity in 1999: "Heuvelmans is equating the creature to what has been reported from Asia," he wrote. "He is only mistaken in thinking they are all Neanderthals. Sanderson was correct in ruling out Neanderthals as the source of the Iceman. It was *Homo erectus*."

A specific feature Hall mentions convinces me of this identity. It relates to a morphological trait that had been described by Sanderson: "A most notable feature of the palmar surface of the hands is one that puzzles us. This is that there is an enormous and prominent pad on the 'heel' at the outer side, behind or 'above' the fifth digit back."

I think this structure found on the original body of the Minnesota Iceman was the result of a well-observed sleeping posture in unknown hairy hominids of Asia. The bent-knees, bent-elbow, face-down on the ground sleeping position of Wildmen was drawn by the Russian V. A. Khakhlov in a 1913 report on the Wildmen submitted to the Russian Imperial Academy of Sciences, as Sanderson noted in his 1961 book. Independently, a Mongolian researcher named Damdin produced a similar sketch of this behavior, but with the hands' palms on the ground, based on sightings of Almas in the 1960s, according to Igor Bourtsev. Hall and I theorized that Almas are *Homo erectus* survivors, not Neanderthals. Thus, by extension, we consider the Minnesota Iceman a *Homo erectus*. For Hall, this feature will prove the reality of the carcass in the future.

I always had my reservations about equating the Iceman with a Neanderthal. In my field guide to unknown hominids, I wrote that the "*erectus*-like features" of the Minnesota Iceman "match quite well some of the reports coming out of Central Asia." The features of the original Minnesota Iceman, if it was an actual physical body, do not seem to be Neanderthaloid. Perhaps it was a Denisovan?

The Denisovan or Denisova hominin is an extinct species first found in March of 2010. Paleoanthropologists announced the discovery of a finger bone fragment of a juvenile female who lived about 41,000 years ago, from the remote Denisova Cave in the Altai Mountains in Siberia. The area is known for contemporary reports of unknown hairy hominids. Furthermore, fossil Neanderthals and prehistoric *Homo sapiens* at one time found their way into this cave. Two teeth belonging to different members of the same population have since been reported. In November 2015, a tooth fossil containing DNA was reportedly found and studied. We don't know what the Denisovans looked like because we have so few physical remains. But their DNA genome has been completely sequenced. The Denisovans are seen as a "cousin of Neanderthals." No telling what their appearance might turn out to be.

The survival of any relict hominid forms, needless to say, would be remarkable. But *Homo erectus* may have survived as late as the known examples of some Neanderthals—30,000 years before present. We now know that the incredible

Australopithecus-like or evolved small *Homo erectus* form, the Hobbits of Flores, *Homo floresiensis*, survived until, at least, 50,000 years ago.

Mark Thomas, an evolutionary geneticist at University College London, said in 2013: "What it begins to suggest is that we're looking at a *Lord of the Rings* type world—that there were many hominid populations." Also in 2013, David Reich of the Harvard Medical School observed: "The best explanation is that the Denisovans interbred with an unidentified species, and picked up some of their DNA. Denisovans harbor ancestry from an unknown archaic population, unrelated to Neanderthals."

Doubts Will Never Cease

The possibility remains that the original Minnesota Iceman was nothing more than a carnival gaff, a created fake, a hoax. The history of the Minnesota Iceman and the manufactured replica is a confusing one. Was Heuvelmans fooled? The debunkers certainly think so.

Heuvelmans was convinced the Minnesota Iceman was a real body. He was certain it was a Neanderthal. But psychologically, Heuvelmans was predisposed to "believe" the Iceman was real; he was not even as skeptical as Sanderson. Ultimately, neither touched the specimen. It remained encased in ice. So, yes, it could have been a fabrication from the beginning.

But there will always be a doubt. The parasites seen on the body, the vegetable matter viewed in the teeth, and the odor from the alleged decomposition—as well as the differences seen between the 1968 and post-April 1969 photographs—all point to the Minnesota Iceman having been an actual carcass.* Maybe it was.

What we do know, as Paul LeBlond has pointed out, is that Heuvelmans' struggle to convince the scientific establishment of the authentic nature of the Minnesota Iceman was a failure. With only a few exceptions, he never managed to stir up the interest of professional anthropologists and paleoanthropologists. The anecdotal nature of the origins of the material evidence of the Iceman, and even Heuvelmans' meticulous analysis of the alleged carcass he examined in 1968, could not overcome the doubtful provenance and inaccessibility of the specimen.

---

*   Herpetologist Terry Cullen recalls his viewing of the Iceman: "There were a number of exoskeletons from parasites (likely some kind of lice) in the body hair. In completely exposed (no ice) areas of the body, one could readily see that each hair had a pore and there were some pores that did not have hair. Other dermal imperfections/irregularities were also present (dark spots, 'bumps,' etc.) that would argue against a construct. There was a significant amount of blood in the ice and on the skin near the areas of injury (skull). Some bone could be identified in the injured area. Regarding the damaged eye, one could observe the expected neural and muscular structures attached or torn. There were what appeared to be vegetative remains lodged in between two of the incisors; the teeth also had mottling and discoloration. Where they could be observed, toe and fingernails were somewhat ragged and had superficial imperfections. The odor of corrupted tissue was certainly present when one got very close to the seams in the glass case. The later (post Sanderson/ Heuvelmans) 'creatures' were obvious fakes that bore little resemblance to the original."

Afterword References

—1969. "Does Neanderthal Lurk in Canada?" Winnipeg, Manitoba: *Free Press*. March 19.

—1985. "Attendance Soars," *Herald and Review*. Decatur, Illinois. August 13, page 7.

—2013. "Mystery Human Species Emerges from Denisovan Genome." *New Scientist*. November 19.

Auel, Jean M. 1980. *The Clan of the Cave Bear*, New York: Crown. *Earth's Children* series.

Bourtsev, Igor. 1982. "The Abominable Snowman: The Riddle Persists," *Asia and Africa Today*, No. 2: 59.

Buhs, Joshua Blu. 2009. *Bigfoot: The Life and Times of a Legend*. Chicago: University of Chicago, p.149.

Busti, Steve. 2013-2016. Personal communication. Austin: Museum of the Weird.

Coleman, Loren. 2001. "Milestones: Bernard Heuvelmans (1916-2001): An Appreciation of a Friend" *The Anomalist* http://www.anomalist.com/milestones/Heuvelmans.html

Coleman, Loren and Patrick Huyghe. 2006. *The Field Guide to Bigfoot and Other Mystery Primates*. New York: Anomalist Books. (A reprint of the 1999 Avon edition.)

Coon, Carleton S. 1962. *The Story of Man*, New York: Alfred Knopf.

Cullen, Terry. 2016. Personal Communications. St. Augustine, Florida: International Cryptozoology Conference, January 5 and April 28.

Culotta, Elizabeth. 2007. "Ancient DNA Reveals Neandertals With Red Hair, Fair Complexions," *Science* 318 (5850): 546–7. October 26.

Doran, Tom C. 2014. "Ag Heritage, Entertainment Highlight Illinois State Fair," *AgriNews*, July 30.

Emery, C. Eugene. 1981-1982. "Sasquatch-Sickle: The Monster, the Model, and the Myth," *Skeptical Enquirer* 6: 2-4, Winter.

Fiacconi, Marta and Chris O. Hunt. 2015. "Pollen taphonomy at Shanidar Cave (Kurdish Iraq): An initial evaluation," *Review of Palaeobotany and Palynology*. December. pp. 87-93.

Gee, Henry. 2004. "Flores, God and Cryptozoology," *Nature*, October 27.

Hall, Mark A. 1999. *Living Fossils: The Survival of Homo Gardarensis, Neanderthal Man, and Homo Erectus*. Bloomington, Minnesota: Mark A. Hall Publications.

Hall, Mark. 1969-2016. Personal communication. Bloomington, Minnesota.

Hansen, Frank. 1970. "I Shot the Ape-Man Creature of Whiteface," *Saga Magazine*, July, 8-11, 55-56, 58, 60.

Heuvelmans, Bernard. 1969a. "Note préliminaire sur un spécimen conservé dans la glace, d'une forme encore inconnue d'Hominidé vivant: Homo pongoïdes (sp. seu subsp. nov.)," *Bulletin of the Royal Institute of Natural Sciences of Belgium*, 45: 4. February 10.

Heuvelmans, Bernard. 1969b. "Neanderthal Man," *Personality*, Bloemfontein, South Africa. June 9.

LeBlond, Paul. 2016. Translator's Notes.

Loof-Wissowa, Helmut. 1996. "Confirmation of the Existence of Neanderthal-like Relic Hominids in Laos: A Preliminary Report." *XIII° Congrès de l'Union Internationale des Sciences Préhistoriques et Protohistoriques*, Foni, Italy.

Lukas, Paul. 1999. "That Fresh Woodstock Feeling," *Fortune*. May 10.

Magraner, Jordi. 1992. *The Relic Hominids of Central Asia*. Valence: Ed. Troglodyte.

Peck, Adam. 2012. Personal communication. Cottage Grove, Minnesota: Zywiec's Landscape and Garden Center. October 3, October 8, October 15.

Peck, Adam. 2013. Personal communication. Cottage Grove, Minnesota: Zywiec's Landscape and Garden Center. February 21, February 24.

Peck, Adam. 2016. Personal communication. April 4.

Prescott, Todd and Mike Quast. 2016. Frank Hansen: Personal Communication to Loren Coleman. March 30.

Rimmer, Lee. 2016. "The 'evolution' of Neanderthals over the last 100 years says more about us" *Ancestry*. February.

Sanderson, Ivan T. 1961. *Abominable Snowman: Legend Come To Life*, Philadelphia: Chilton.

Sanderson, Ivan T. 1969a. "The Missing Link," *Argosy Magazine*. May, 23-31.

Sanderson, Ivan T. 1969b. "Preliminary Description of the External Morphology of What Appeared to be the Fresh Corpse of a Hitherto Unknown Form of Living Hominid," *Genus* (Rome) 25 (1-4): 249-278.

Sanderson, Ivan T. 1970. "Bozo: The Iceman," *Pursuit*, October.

Sanderson, Ivan T. 1972. Personal communication to Tom Hall, *The Sunday Times*. January.

Shackley, Myra. 1983. *Still Living? Yeti, Sasquatch and the Neanderthal Enigma*. New York: Thames and Hudson.

Spavin, Don. 1969. "Is Frozen Mystery 'Man' Just a Hoax?" *Pioneer Press*, St. Paul, Minnesota. Sect 2: 8. March 30.

Solecki, Ralph. 1971. *Shanidar: The First Flower People*, New York: Alfred Knopf.

Stringer, Christopher and Clive Gamble. 1993. *In Search of the Neanderthals*. New York: Thames and Hudson.

Sykes, Bryan. 2016. *Bigfoot, Yeti, and the Last Neanderthal: A Geneticist's Search for Modern Apemen*. Newburyport, MA: Disinformation/Red Wheel/Weiser, LLC.

Walia, Arjun. 2015. "New Ancient 'Mystery Human' Species Identified Shakes Up The Theory Of Evolution," *Collective-Evolution*, March 1.

West, Rick. 2012. Personal communication. Green Bay, Wisconsin: Dr. West's Sideshows. February 24.

Wolff, Mike. 1969. "Prehistoric Man Still Living? Exhibit Stirs Controversy," *Star*, Minneapolis, Minnesota. March 20.

Appendix A

Bernard Heuvelmans' Original Introduction
to *Neanderthal Man is Still Alive*

ON THE TRACK OF UNKNOWN "MEN"
From cryptozoology to the search for our wild brethren

"You might also wish to know where I was born,
for today one believes that where a child has
uttered it first cries is an essential element of its
nobility." — Erasmus, *In Praise of Folly*

How did I come to study animals, and from the study of animals known to science, how did I go on to that of still undiscovered animals, and finally, more specifically to that of unknown humans?

It's a long story.

For me, everything started a long time ago, so long ago that I couldn't say exactly when. Of course, it happened gradually. Actually—I have said this often—one is born a zoologist, one does not become one. However, for the discipline to which I finally ended up fully devoting myself, it's different: one *becomes* a cryptozoologist. Let's specify right now that while cryptozoology is, etymologically, "the science of hidden animals," it is in practice the study and research of animal species whose existence, for lack of a specimen or of sufficient anatomical fragments, has not been officially recognized.

I should clarify what I mean when I say "one is born a zoologist." Such a congenital vocation would imply some genetic process, as in a lineage of musicians or mathematicians. But there was nothing hereditary in my becoming a zoologist. On my father's side, I stem from lawyers and philologists; on my mother's, from a constellation of actors, painters, and musicians; so I'm a happy marriage of arts and reason, of rigor and sensitivity, and up to a point, of deductive work and scientific detective inquiry required by any judicial issue, with the intuition required for its resolution. That might explain how from being a zoologist I became a cryptozoologist, up to the point of being dubbed, the "Sherlock Holmes of Zoology."

If I think I was born a zoologist, it is because even when probing deep in my memories, I can't discover a beginning for my passion for animals. It seems to me that it has always been so. That said, my earliest striking recollection—the only one perhaps from the first year of my life—involved a large fish stranded on the pebbles of Trouville; it couldn't have been longer than a meter, but to my childish eyes it seemed enormous. Who is to say that wasn't the spark of my interest in "sea monsters" to which I have devoted two books? As to my first toy, besides the traditional teddy bear, it was a small untearable canvas book of an animal alphabet. Being in English, it went from A for aardvark to Z for zebra. Perhaps it was not such a coincidence that my doctoral thesis focused on the dentition of the aardvark and that later some zoologists ended up thinking of me a "drôle de zèbre."* An "odd bird," indeed, who has never accepted with eyes closed what he was told or ordered to believe.

All told, it would be more appropriate to say that I am a zoologist because I have been subject at an early age to an *imprinting*. Just like those baby geese of Konrad Lorenz that followed the first living being they saw after hatching, taking it for their mother, I was drawn in by the animals of my first alphabet book, believing them to be close relatives. That would certainly explain the deep love I have for them, and the fact that I feel the most perfect connection with those said to be my remote kin. Undoubtedly, this could also shed some light on the nature of my main concern, which has haunted me from the beginning of my studies: the problem of the origin of Mankind.

How I later became a cryptozoologist is easier to explain, perhaps as I said, because of my heredity.

Let's skip quickly over those rather classic episodes of the career of every budding zoologist. As for me, they went from hunting butterflies to raising ladybugs, spiders, and white mice; from the transformation of kitchen sinks into aquaria for tritons, sticklebacks, and diving beetles; to welcoming under the family roof all needy animals, be they rats or hedgehogs, garter snakes or swifts, or even stray dogs; from excursions through the woods and the dunes of the seashore to the daily visits to that marvelous zoo in Antwerp, where I spent my holidays with my maternal grandparents. That faithful attendance that usually focused on the monkeys' cage had long-term repercussions. My first concern after I got married and had a place of my own was to acquire a capuchin monkey, something that I had dreamed of since I was a child, but this was clearly not the playmate that normal parents would choose for their child when they lived in an apartment. That little monkey was the first of a series of simian children and friends that have brightened my life: I see in them a reminder of Paradise Lost, and they have taught me real wisdom.

This passion was maintained from its earliest days by an enthusiasm for the whole of the zoological literature, available first in the communal library and then in that of the college. At the age of twelve, I had already read Fabre and Buffon, Cuvier and Darwin. As a distraction from the more difficult parts of their works, I read stories about Red Indians, hunting stories (which prematurely made me hate all killers), as well as the charming popular works of naturalist Henri Coupin, particularly the one entitled *Les Animaux Excentriques*. That book clearly was the trigger for my

---

*   A French expression meaning literally "strange zebra," or colloquially "odd bird." —The Editor

later investigations; having reread it recently, I was pleased and surprised to find in it all those extraordinary creatures that appear today in my own books, from the gigantic octopus and the sea serpents to moas and dodos, pterodactyls and dinosaurs, megatheriums and Pithecanthropes.

During my studies at the school run by Jesuits, I often put aside the study of Latin, Greek, and apologetics to enjoy the fruits of the Tree of Science, and my tastes in literature made me prefer *Extraordinary Stories*, a collection of tales by Edgar Allan Poe, to those of Alphonse Daudet in *Lettres de Mon Moulin*, Voltaire to Bossuet, Jules Verne to Madame de Sévigny, and Sherlock Holmes to Ruy Blas or d'Artagnan. My real heroes were not called Napoleon or Joan of Arc, but Marco Polo and Humboldt; Paul du Chaillu, the Homer of gorillas; and Robert F. Scott, the martyr of Antarctica. And to Saint Francis of Assisi, who humbly communed with the lowest animals, I greatly preferred Tarzan, who was one with the animals, and sorely grieved at the tragic death of the she-ape who had nursed and raised him.

For amusement, in literature as well as in cinema, I clearly opted for crime stories and exotic adventure stories. Three novels of my youth have played a significant role in the gestation of cryptozoology: *Twenty Thousand Leagues Under the Sea* by Jules Verne, which opened the doors to the secrets of the sea, *Les Dieux Rouges* (*The Red Gods*) by Jean d'Esme, whose work evoked the reality of ape-men in Indochina today, and, above all, Conan Doyle's *Lost World,* which imagines the survival of the varied fauna from past ages on an isolated high plateau in South America.

The background was in place, the climate well established, and the tragicomedy was about to begin.

My doctoral thesis in science followed in the orthodox tradition of classical zoology. It dealt with, as I mentioned, the dentition of the aardvark. Of course, some whispered that having chosen to study the teeth of an animal that was not supposed to have any, being an edentate, showed a perverse taste for paradox, a sense of humor that has no place in science. Actually, this choice was for me almost natural, perhaps unavoidable. I wanted to specialize in the study of mammals, my kin. Among mammals, dentition is the most characteristic feature, the first that should be well understood: the shape of the teeth plays the same role in the identification of mammal genera as fingerprints do in the identification of individuals of our species. And, among the teeth of mammals, the most mysterious, the most incomprehensible, the most difficult to classify, were those of the aardvark, the "earth-pig" of the Transvaal Boers, a strange, digging, termite-eating ungulate. Formerly, it used to be classified among the Edentata, a group that brought together all terrestrial mammals without anterior teeth or having no teeth at all, or having in any case only atrophied teeth, without enamel. But rest assured: the aardvark does have teeth, although they are most bizarre; they are like stacks of innumerable little tubes, hexagonal ivory prisms resembling the alveolae of a bee hive. They have even been compared to the teeth of some fossil rays! This dentition would clearly be the focus of my research. I am already known for my fascination with enigmas, but not because I enjoy the fog of mystery or the shadows of the unexplained. On the contrary, I consider an enigma a challenge to be met. Mysteries attract me because they provoke the urge to solve them.

It took me more than two years to solve the problem of the aardvark's dentition. I studied all the skulls available in France and in Belgium and dissected an intact head that my mentor, Dr. Serge Frechkop, had the good fortune of finding in the Congo. I conducted a series of longitudinal and transversal sections of teeth and of whole mandibles. What I found—which I had already suspected from the position of the aardvark within the family tree of ungulates—is that its dental tubes were simply the sheaths of the extremely subdivided fingers of the pulp. In other words, the aardvark's tooth was one in which the cusps of the crown were highly multiplied. For example, the hippopotamus' third molar has 4 cusps; the warthog has around 25; some Asian elephants have up to 90 and aardvarks have hundreds (up to 1,500 in the second molar). The aardvark's tooth, in spite of its appearance, thus fits within the usual pattern of mammalian teeth, of which it is an extreme case. From a tooth that was regarded as mysterious, thought perhaps to have a non-mammalian origin, I succeeded in understanding as a normal tooth, one that could have evolved from the simpler teeth of other mammals. The "monster" had been tamed.

If I have written at length about a problem that seems remote from that of the hairy wild men, it's because it is characteristic of my way of thinking. Using as a pretext that I venture off the beaten path in areas haunted by fantastic creatures and legendary beings, some people label me as "wacky" and pre-occupied with matters of little interest. If they had a sense of humor, they would say that "Heuvelmans is that zoologist who wrote his doctoral thesis about the teeth of an Edentate and has now become a specialist in the study of non-existent animals." (Like Cyrano de Bergerac, I can make my own jokes about my scientific "sense.") I would answer that the unknown, which does not exist at least in textbooks, is always confusing, somewhat disquieting, even terrifying—in a word, fantastical. It is the privilege of innovators to face the Unknown. Should we be surprised that, in his time, Darwin was criticized for having tried to explain the world of animals on the basis of Ovid's *Metamorphoses*? Let us never forget that myths are perhaps the reflection of obscure and misunderstood realities, and that in any case, they are patterns pre-existing in our mind (Carl Jung's archetypes of the collective unconscious?) into which we try to fit, willy-nilly, facts that belong to Science. It is indeed correct that Darwin only proposed a rational explanation for the ancient myth of avatars, a phenomenon the poet (Ovid) grasped only vaguely and interpreted through his imagination. Similarly, I am not after sea serpents to exorcise an antique incarnation of the Demon of External Darkness, but rather matter-of-factly to try to discover if the real animals behind this myth are to be classified as Fish, Reptiles, or Mammals, which parts of the ocean they inhabit, what is their mating season, what they eat, whether they are gregarious or solitary, etc.

We are finally approaching, through a wide centripetal spiral, the very topic of this volume.

Throughout my studies, I acquired the habit, shared I believe with many of my colleagues, of storing in a special file, as much by interest as for amusement, newspaper clippings and bibliographic references about strange or unexplained facts about the world of animals that might one day become a subject of research. Within this grab-all, there were sensational articles about the Loch Ness monster, serious psychological studies on wolf-children, stories about mass strandings of giant squids on some remote beach, live toads found within old rocks, showers of fish, the kidnapping of black women by salacious gorillas, mammoths glimpsed in the Siberian taïga, a strange bear terrorizing East Africa, midget Pithecanthropes in Sumatra, and even a dragon hunt in the Swiss Alps. There were also photos of an ape-man supposedly discovered in the Atlas Mountains, a scaly rhinoceros shot down in Sumatra, and a legged fish said to have survived from the Devonian era. There were also articles, especially related to that last one—the famous coelacanth—discussing the scope and wealth of zoological discoveries still to be made. Just to mention really new large animals, since the day of my birth the current existence of the following had been established: the Congolese aquatic civet, the freshwater dolphin of Tung-Ting Lake, the pygmy chimpanzee, the Congolese peacock, the kouprey or grey Cambodian bull, and the golden langur. After seventy years of effort, the first live specimen of the giant panda had been captured, many new beaked whales had been discovered, and the coelacanth had been fished out of the relatively shallow waters where it had been hiding since the dawn of time.

Within that motley collection, there was much that could be cast aside. I must admit, since it is not easy to rid oneself of the burden of an education peppered with dogmas and preconceived ideas, that there was not much that I kept. It was with a strange sense of make-believe that I saw early in 1948 an article in the *Saturday Evening Post* entitled "There Could Be Dinosaurs." At first, I didn't even think of clipping it out to include in my special archives; it looked too much like a science fiction story cleverly presented as an authentic document. What led me to hesitate was the name of the author: Ivan T. Sanderson. I knew that he was a well-known naturalist who had led the Percy-Sladen expedition in Cameroon and had contributed a plethora of new species to the natural history department of the British Museum. I had also read his fascinating book on rare animals of the African jungle, *Animal Treasure*. Was it possible that he might also be an author of science fiction stories? (Actually, Sanderson wrote a number of novels for young readers.) To find out, I read the article carefully and made marginal notes. Among the experts quoted was

Carl Hagenbeck, the director of the Hamburg Zoo, as well as its principal purveyor, Joseph Menges, professionals unlikely to engage in tall tales. I researched the sources and even added to them along the way in a process that lasted a few years. It was worth the effort. The result was clear: Sanderson had invented nothing. His work was based on information gathered by prominent people, such as Sir Henry Johnston, the governor of Uganda who had contributed to the discovery of the okapi; the German explorer Hans Schomburgk; and the English naturalist John G. Millais. As incredible as it seems, one could legitimately wonder whether there might exit in the heart of Africa survivors of the great aquatic dinosaurs of the Secondary Era, or at least animals that closely resembled them.

This was such a big deal that I decided to gather in one volume, fully documented and quoting all its sources, the scattered information available on the existence of large animals still unknown to Science. That work took me four years and led to the publication in 1955 of *Sur la Piste des Bêtes Ignorées* (*On the Track of Unknown Animals*), which was widely translated. The small book that I had planned had grown into a two-volume opus. Even then, I had to omit the part, still incomplete, dealing with "unknown animals" of the sea. Exhaustively researched and completed, that part gave birth successively to *Dans le Sillage des Monstres Marins: Le Kraken et le Poulpe Colossal* in 1958 and then to *Le Grand Serpent de Mer* in 1965, which were published jointly in English as *In the Wake of the Sea-Serpents* (1968). And when comes the day when I will publish everything I now have in my files on animals unknown to science, I will surely need ten volumes: that's ignorance that takes a lot of room.

Thus, over about twenty years, a new science, which I soon came to name Cryptozoology, was born and matured.[69] The methodology—a system of study and systematic research—emerged from my tentative first steps and opened the door to powerful methods of screening the evidence. It was a desperate effort, because at the rate at which the world fauna shrinks and species disappear one after the other, like unripe fruits blown away by the technocratic hurricane, one had to act quickly. Even if the only fruit of this methodology were to be the upsetting revelation that is the topic of this book, I believe the cryptozoological adventure would have been worthwhile.

One additional detail, but an important one: this Ivan Sanderson, the author of the article in the *Saturday Evening Post*, which had prompted me to write *On the Track of Unknown Animals,* was the same Ivan T. Sanderson who would be at the center of the discovery of the specimen that crowned all my theoretical research. He will have

---

69   The first mention in print of this word, which I was already using routinely in my private correspondence, dates from 1959. It appears in *Géographie Cynégétique du Monde*, the work of my colleague and faithful friend Lucien Blancou, who honored me by dedicating it to me as "Master of Cryptozoology." [The first mention in English was by Ivan T. Sanderson (1961).—The Editor.]

been the alpha and the omega of this saga.

The current episode is, of course, only one among many in cryptozoological research, but it is undoubtedly today the most important, because on the one hand it pertains to the fundamental problem of our own origins, and on the other because it was settled by the careful and detailed examination of a concrete specimen and thus received a genuine zoological baptism. *Homo pongoides* is, in a way, the proof of the effectiveness of the cryptozoological method.

How did an eminent Soviet historian and philosopher, who was essentially concerned with discovering the anatomical, physiological, ecological, and social conditions that led to the emergence of *Homo sapiens* ever get involved in this cryptozoological adventure? How did the in-depth research of Professor Boris Porshnev in the USSR, combined with the work of Sanderson in the USA, and mine in France manage to achieve such success? That's what I shall now describe.

The original edition of my book *Sur la Piste des Betes Ignorées* (1955) included a number of chapters dealing with man-like unknown creatures. A whole section of the book was indeed entitled "The human-faced beasts of Indo-Malaysia," where I reviewed the problem of the Nittaewo, the hairy Ceylonese midgets exterminated around 1800, as well as that of the similar Orang Pendek apparently still present in Sumatra. I also mentioned phantasmagoric gnomes from Indochina, with a tail and forearms as sharp as cleavers; I also wrote of hairy men of normal stature and appearance recently observed on Christmas 1953 in the Malacca Peninsula. Finally, I dealt at length with the mystery of the Himalayan Snowman, which was much in the news in the 1920s and back in the limelight in the 1950s; no alpinist could venture to climb Everest without finding its footsteps in the snow. I rebelled against its ridiculous nickname, based on an error in translation; in my mind, there was nothing Abominable in the Snowman, it was not a man and did not live in the snows. Besides that, I had devoted two complete chapters to similar hominid and hairy creatures seen in other parts of the world, including an anthropoidal great ape found in South America and furry dwarves from East Africa, variously named Agogwe or Mau.

Following the flood of correspondence after the publication of my book, and given the continuing progress of my bibliographic research, I was able to significantly enhance the information contained in these chapters in the English edition which appeared in 1958 as *On the Track of Unknown Animals*, and even more so for the extended English edition of 1962. In the meantime, it occurred to me that the problem of the Snowman was much more complicated than imagined, and that it encompassed three kinds of rather different creatures. There were also rumors about the existence of wild hairy men throughout South America, from Colombia to the Guyanas, in northern Chile, in Bolivia, in northern Argentina, and in various states of Brazil. Finally, in Africa, rumors of hairy dwarves came not only from the southeast side of the continent, but also on the other side, in the Ivory Coast. I was even soon to receive additional information about the presence of tall "hairy men" from the Congo, which was still Belgian at that time.

In my book I did not think for a moment of identifying all these creatures scattered over most of the planet, which actually had only two traits in common: that of looking more or less human, and that of being also very hairy. There were too many differences between them: some were minuscule gnomes, others were the size of an average human, and still other were real giants; some had a receding brow, others a skull like a brick; some were always bipedal on flat ground, others sometimes ran on all fours; some left small triangular footprints, others large prints where the big toe was well to the side, and others left huge prints, long and narrow, with very long toes; finally—and this was of little importance—some were reddish, others puce, and some gray or even deep black.

I commented on each of these types in isolation, within their regional context, and merely expressed some speculative hypotheses on the nature of those best described. The famous Ameranthropoid of the Columbia and Venezuelan hinterland appeared to me to be a cebid monkey (a family which includes all American monkeys except marmosets and tamarins) having perhaps reached through convergent evolution a form and a stature comparable to those of the great apes of the Old World. I was wondering whether the small Agogwe of Tanzania and Mozambique might not be relic Australopithecines. I agreed with my eminent colleague (and later friend) Dr. W.C. Osman Hill in his hypothesis (expressed as early as 1945) that the Sri Lankan Nittaewo and the Sumatran Orang Pendek might be related to the extinct Javanese Pithecanthrope. I expressed for the first time the idea that the larger Himalayan Snowman was likely to be related to *Gigantopithecus*, a giant anthropoid ape of the mid-Pleistocene from the Chinese province of Kwangsi, and I supposed that the others were relics of the fossil Siwalik fauna, which was so rich in monkeys. Not for an instant did the idea that one of these "hairy men" might be a real human cross my mind: for all we knew, they lived like wild animals, had no articulated language, and did not use tools or fire. At best, some had the reputation of throwing stones or fighting with sticks, which even chimpanzees do. I had actually deliberately left out of my files all creatures that clearly seemed human, such as the Maricoxis encountered by Colonel Fawcett in the southwest of the Matto Grosso; although they were strangely hairy for Indians, they used bows and arrows.

Ivan T. Sanderson, with whom I had begun to correspond regularly in 1957, generally agreed with my views. It was he who, as early as 1950, had first thought of the Himalayan Snowman as a survivor of the old simian fauna of the Siwaliks.

However, the whole perspective shifted when Professor Boris F. Porshnev stepped in. He relates in the first part of the French work *Neanderthal Man is Still Alive*, that in January 1958, his interest suddenly focused on the Snowman following the sighting of one in the Pamir. So when the highly condensed Russian translation of my book appeared in Moscow, Porshnev, who was very familiar with French and a prominent Soviet expert on the history of France, was eager to consult the original version of the book. He quickly got in touch with me. That was the beginning of a close friendship and fruitful collaboration, which substantially enriched my knowledge of the subject. In 1961, he came for the first time to visit me in Paris where, as the newspaper say— and it was true—we had a frank and fruitful exchange of views and survey of the problem. Two years later he published the admirable monograph *The Present State of the Problem of Relic Hominoids*, expressing his personal views of the situation.

One must say that Porshnev and I were both quite convinced of the existence of wild hairy men in the broadest and loosest sense. However, there was a point on which we never managed to agree: I saw the Himalayan Snowman as an anthropoid ape, while Porshnev saw an actual human, more specifically a Neanderthal man survivor from the recent Pleistocene.

There was a good reason for this difference of opinion. For my part, I had researched the reports of western travelers—British, French, Austrians, Swiss, Italians, Americans, etc.—most of them alpinists approaching the Himalayan peaks from the southwestern side of the mountain range, as well as the reports from their informers from Nepal, Sikkim, Bhutan, and Kashmir. As for Porshnev, he had focused his attention on reports originating from areas on the northeast side of the Himalayan range: both from states of the Soviet Union—Tadjikistan, Kirghizstan, Kazakhstan, and the Siberian vastness—and from Tibet, Sinkiang, Outer Mongolia, and China, and finally the Caucasus, on the doorstep of Europe.

On the south side of the mountain range, which divides Asia in two distinct zoogeographic zones (the Palearctic and the Oriental), the focus of my own research was on what seemed to be a strangely bipedal monkey (of course, so is the gibbon). Its squarish head suggested the presence of a sagittal crest, like that of the gorilla, or perhaps the existence of a standing tuft of hair as on the coconut-shaped head of young orangutans. The Yeti, as it was called in Nepal, was also said to run on all fours when in a hurry.[70] Finally, a whole series of behaviors—scratching, flashing teeth as a means of intimidation, a perverse taste for destruction, the manifestation of a powerless rage by bouncing up and down on the spot while pulling off tufts of grass—emphatically suggested an ape. I have observed many and even raised some. However, north of the Himalayas, Porshnev observed in his Snowman characteristics that suggested a real human: body proportions, a non-opposable big toe, long head-hair, and consistent bipedalism (except of course when climbing steep slopes).

---

70   Perhaps, on the other hand, the Yeti might be bipedal only when it crosses the snowfields where its footprints are found. The American anthropologist Sidney Britton observed a chimpanzee who stood on its hind legs to walk across snow. It is evident that for an anthropoid ape, that is the best way to minimize the skin area in contact with the icy substrate and hence also minimizing the cooling of the whole body.

We could not agree because we were obviously not talking about the same thing. That's what I finally understood. In the Himalayan region, I had already tried to show in my book that three distinct types of creatures were confounded, perhaps due to legend. The creature studies by Porshnev seemed to point to a fourth type, clearly human, which the Soviet scientist had excellent reasons to consider as Neanderthal.

But that's not the end of the story! Things got even more complicated in 1958. At that time, Ivan Sanderson had undertaken an extensive journey through North America to document his work *The Continent We Live On*. He was already on his way when many of his correspondents, of which I was one, sent him newspaper clippings mentioning the discovery in the Klamath Mountains of northern California of absolutely enormous human footsteps. This incident was related to an ancient enigma that had long attracted the attention of Canadian researchers like J.W. Burns and René Dahinden. For centuries, the Indians of British Columbia claimed that hairy giants, which they called *Sasquatch,* lived in the Rocky Mountains; various sightings by Pale-Faces had confirmed it. Sanderson left to conduct an inquiry on the newly discovered footprints; journalist John Green, in Agassiz, BC, began to devote much of his time to this problem and was quickly followed by a bevy of enthusiastic amateurs. A new saga was launched—that of Bigfoot.

The oversized tracks made by that giant didn't resemble those of the Yeti. However, the press soon dubbed the creature "the American Abominable Snowman"! Thus arose more confusion, which, alas, Porshnev soon supported.

Nevertheless, when in 1961 Sanderson finally published his expected vast synthesis of the problem, *Abominable Snowmen: Legend Comes to Life,* he wisely proposed to classify the various types of hairy hominoids seen on five continents (or rather zoogeographic zones) in four categories: (1) *Sub-humans* (i.e. Neanderthals), (2) *Proto-pygmies*, (3) *Neo-giants*, (4*) Sub-hominids* (i.e. anthropoid apes).

In the mind of the Scottish-American naturalist, the *Sub-humans* were the Almas and other hairy Asian wild men researched by Porshnev. The Proto-pygmies included the Orang Pendek from Sumatra; the Teh-Ima, the smallest of the Himalayan snowmen; as well as the African Agogwe and the Duendes of tropical America. The Neo-Giants were represented by the Sasquatch, aka Bigfoot, and the largest Snowman, Dzu-The. And finally the *Sub-hominids* were represented by the Mi-Teh, the mid-sized snowman. Some of Sanderson's attributions were arguable, but he certainly had the merit of clarifying a business that was wallowing in confusion.

In contrast to the splitting process that Ivan and I adopted, Boris Fedorovitch lumped together the various types of hairy bipeds found here and there on our planet. To him, they were all small relic populations of Neanderthals, diversified by their adaptation to different local conditions.

I often pointed out to him during our conversations and in our correspondence that there were among the various types traits that were difficult to reconcile with the idea of a single species. Differences in hair color were of no consequence; after all, Senegalese, Swedes, Mongols, and Bushmen all belong to the same species. Even differences in stature could be explained; in central Africa, Wambutti pygmies and giant Watutsi are very different neighbors but are nevertheless just as much *Homo sapiens*. One might even be able to explain, by reason of age or sexual dimorphism, some differences in color and stature as well as some diversity in structure. For example, it once took a long time to realize that all described species of orangutans boiled down to three types: large adult males, much smaller females, and immature juveniles. However, there were among the wild hairy men some traits that implied at least some specificity. Individuals with a square head could not belong to the same species as those of the same gender with a sloping forehead and a flat head. Individuals from the same species could not sometimes leave footprints longer and narrower than those of modern man, and other times shorter and wider, and sometimes plainly triangular.

My friend Porshnev answered these points with a variety of arguments. First of all, he said, we are not measuring the magnitude of individual and racial variations of the species and we have no knowledge of the workings of its foot. He added that we already had our work cut out to convince people of the existence of a single species, and that we would never manage to achieve acceptance of a pleiad of hairy savages. He thought that until we had more information, it would be wise to suppose, as a working hypothesis, that there was only one species, a relic Neanderthal.

To which I answered that such wisdom might eventually turn out to be a folly we would sorely regret. As a matter of fact, hairy wild men had been mentioned since time immemorial. Both the Bible and the Babylonian Epic of Gilgamesh mention them. Classic antiquity had its Satyrs and Silvans, and the Middle Ages their Wudéwasa and Men-of-the-Woods. From the Renaissance on, in a period of progressive exploration of the world, travelers had seen them in many areas. To the point that Linnaeus himself had included them in his *Systema Naturae*. It was the discovery of the great apes, first the chimpanzee and the orangutan, and much later the gorilla, that was to relegate them to legend. Actually, all the great apes had originally been described as "wild men" or "hairy men" or "men of the woods." With these discoveries, nineteenth century science thought that it had explained once and for all a thousand-year-old enigma. Henceforth, all new stories of wild hairy men were thought to be the result of errors or hoaxes.

One had to avoid a repeat of such a mistake. If the first specimen to be discovered was to be identified as Neanderthal, the problem would be considered solved and any subsequent attempt to search for a living *Gigantopithecus*, Pithecanthrope, *Australopithecus*, or any unknown anthropoid ape would be considered a folly. And if, on the other hand, the first specimen captured and identified turned out to be an unknown species of ape, the hypothesis of the survival of Neanderthals would be subject to ever-lasting ridicule.

Furthermore, I thought that from the very beginning we should adopt the attitude of scientists, not of diplomats. Diplomacy, often based on white lies, cannot co-exist with Science, which is entirely devoted to the search for Truth.

Anyway, Porshnev was not absolutely opposed to the idea of there being many species of wild men. But he was actually interested in only one of them. There was something of a preconceived idea in his approach, which he readily admitted and which is clear in his part of the book: his interest was uniquely the problem of the origin of Man, and if the object of our research had not been Neanderthal, he would have had no interest at all in the question. Although the problem of anthropogenesis was also important to me, I could not as a cryptozoologist adopt such an attitude.

If, from my point of view, Porshnev's approach to the problem was somewhat misguided, one must recognize that in identifying one type of hairy biped with Neanderthal Man, he had taken the first step ahead of the rest of us: he showed remarkable flair, one of those flashes of genius at the root of scientific revolutions.

Because, finally, his hypothesis had everything going against it. Wherever they were seen, hairy wild men were always described as real animals, living like animals, without any of the features that distinguish Man: articulated language, tool-making, the use of fire, and organized social life.

Did Neanderthals have language? On this matter, anthropologists differ. We have good reasons to believe that these prehistoric men had a rather rough stone-tool making industry, showing a high degree of manual dexterity, that they used torches, that they sometimes lived in communities involving many families, that they used cosmetics and wore jewels, that they buried their dead with grave offerings, and that they even had a cult of the Bear.

That doesn't really jive with the image of a man-beast. That is why I have long had a strong aversion to my Soviet friend's hypothesis. It took the full deployment of a brilliant demonstration of his theory to shake my aversion. And I finally bowed to the evidence when I had under my very own eyes the irrefutable proof of the current existence of creatures anatomically similar, in the smallest details, to the Neanderthal of long ago: a skin and bone specimen.

In *Neanderthal Man is Still Alive*, we first hear Professor Boris F. Porshnev's broad autobiographical account of the question on which he spent most of his efforts over fourteen years of his life. In the second part of the book, I relate how Ivan Sanderson and I came to examine the perfectly preserved corpse of one of these creatures, which I then spent years studying.

The reader will first benefit from a panoramic view of the question before being invited to a closer examination. He should not succumb to the temptation of leaping over the first part to immediately attack the second under the pretext that it is concerned with a tangible and measurable anatomical specimen rather than with legends, hazy testimonies, and equivocal footprints. To insist on a specimen would be a grievous mistake that many people make when dealing with this kind of investigation. A specimen has only subsidiary value, as an *a posteriori* verification. It is actually impossible to understand and appreciate the identification and study of an individual without having first assimilated the complex background of the problem.

Before passing the microphone to my friend Boris, I would like to emphasize another point, a practical rather than theoretical one with strong moral implications, where I cannot agree with him, nor with Sanderson. They both advocate, as the next step in our study to seek a new specimen, shooting one down. I strongly object to such an attitude. Porshnev justifies it by claiming that the creatures that we are looking for are not *really* human, at least in the philosophical sense of the term. Whether that is so or not is quite irrelevant in my view.

We have no moral right to dispose of the life of other creatures beyond the necessity to feed ourselves; that is a stark law of Nature and we are certainly not, alas, herbivores. We have even less the right to kill when dealing with intelligent and feeling beings, capable of emotions and suffering. We even have a duty to protect them, as they are rare and menaced with extinction. The loftiest demands of Science can never justify murder or torture. One may also doubt whether killing one of these creatures would appreciably benefit our knowledge. George Schaller has taught us much more about gorillas by observing them for a year in their natural habitat than did a century of massacres and captures often taking place in atrocious conditions.

To believe that possession of a specimen, an "irrefutable proof "of existence, could convince the scientific world of the existence of such creatures is a mark of great naivety and ignorance of the history of zoology and particularly of anthropology. The second part of the book will show once again that when a sample embarrasses Science, and does not fit within the scope of traditionally accepted facts, the representatives of Science do not even bother to examine it. Even if they had to, they would immediately claim that it must be a clever hoax or an abnormal individual. Public and scientific opinion does not seem ready to receive such an upsetting revelation as the existence on our planet of another form of humans.

Thus, the best we can do at this stage is to inform the broad public as well as scientists, to make them aware and to prepare them to accept the situation. That is even for Porshnev and myself an imperative duty. Brecht essentially said it in one of his plays: "He who knows nothing and says nothing is only ignorant, but he who knows and says nothing is a criminal."

Appendix B

TABLE OF MEASUREMENTS OF PONGOID MAN (in cm)

Height
a) With knees naturally bent (GF)	180	+/- 2
b) With legs fully extended (ET)	184	+/- 2

Head
Maximum width	19	+/- 0.5
Maximum bi-zygomatic (cheek-bone to cheek bone)	18.5	+/- 0.5
Bi-goniacal width (width of teeth)	11	+/- 0.5
Maximum length	26	+/- 1
Total height: gnathion-vertex (chin to crown)	25	+/- 0.5
Height of head: tragion-vertex (ear to crown)	14	+/- 0.5
Height of face: nasion to gnathion (nose to chin)	15	+/- 0.5
Width of nose	6	+/- 0.2
Height of nose	5	+/- 0.2
Width of mouth	8.5	+/- 0.2
Inner eye corners distance	5.5	+/- 0.2
Pupil to pupil distance	9.25	+/- 0.2
Outer eye corners distance	12.5	+/- 0.2
Eye width	3.5	+/- 0.2

Trunk
Frontal height (suprasternal notch – pubic symphysis)	62	+/- 2
Height of thorax (suprasternal to xiphoid process)	24	+/- 1
Height of abdomen (xiphoid –pubic symphysis)	38	+/- 1
Bi-acromial (shoulder) width	46	+/- 1
Bo-cretal width	37	+/- 1
Width of thorax	42	+/- 1
Depth of thorax	37	+/- 1

Limbs
Total length of upper limb	88	+/- 2
Length of upper limb except for hand	62	+/- 2
Upper arm (acromial-radial) length	35	+/- 1
Forearm (radial to style) length	27	+/- 1
Length of hand	26	+/- 0.5
Width of hand	12	+/- 0.5
Total length of lower limb (ET)	94	+/- 2
Total length of lower limb (ET) without foot	88	+/- 2
Length of lower limb (GF)		
a) at the ilio-spinal point	90	+/- 2
b) at the trochanterian point	84	+/- 2
c) at the pubic symphysis	81	+/- 2
Length of thigh (iliospinal –tibial)	51	+/- 1
Length of leg (tibial to malleolar)	37	+/- 1
Length of foot	26	+/- 1
Width of foot	16	+/- 0.5

TABLE OF ANTHROPOMETRIC INDICES OF PONGOID MAN

Type	Calculation	Interpretation with respect to *Homo sapiens*
Cephalic	$\dfrac{100 \times 19 +/- 0.5}{26 +/- 1} = 61.1$ to 78.0	Mesocephalic to ultra-dolichocephalic
Height to length of head	$\dfrac{100 \times 14 +/- 0.5}{26 +/- 1} = 50$ to 58.0	Flat head (chamaecephalic)
Height to width of head	$\dfrac{100 \times 14 +/- 0.5}{19 +/- 0.5} = 69.2$ to 78.3	Very flat head (tapeinocephalic)
Nasal	$\dfrac{100 \times 6}{5} = 120$	Extreme wide nose
Inter-orbital	$\dfrac{100 \times 5.5}{18.5} = 29.7$	Eyes extremely far apart
Inter-pupillary	$\dfrac{100 \times 9.5}{18.5} = 50$	-
Zygo-mandibular	$\dfrac{100 \times 11 +/-0.5}{18.5 +/- 0.5} = 55.0$ to 63.8	Abnormally narrow jaws
Height of trunk (GF)	$\dfrac{100 \times 62 +/- 2}{180 +/-2} = 32.9$ to 35.9	Very long to abnormally long
Height of trunk (ET)	$\dfrac{100 \times 62 +/-2}{184 +/- 2} = 32.2$ to 35.1	-
Shoulder width (GF)	$\dfrac{100 \times 46 +/- 1}{80 +/- 2} = 24.7$ to 26.4	Very wide to abnormally wide
Shoulder width (ET)	$\dfrac{100 \times 46 +/- 1}{184 +/- 2} = 24.1$ to 25.8	Wide to very wide
Thoracic	$\dfrac{100 \times 42 +/- 1}{37 +/-2} = 105$ to 122	Nearly cylindrical thorax with deep chest.
Width of basin (GF)	$\dfrac{100 \times 37 +/-1}{180 +/- 2} = 19.7$ to 21.2	Wide to very wide basin
Width of basin (ET)	$\dfrac{100 \times 37 +/- 1}{184 +/-2} = 19.3$ to 20.8	"
Length of upper limb (GF)	$\dfrac{100 \times 88 +/- 2}{180 +/- 2} = 47.2$ to 50.5	Long to very long
Length of upper limb (ET)	$\dfrac{100 \times 88 +/- 2}{184 =+/-2} = 46.2$ to 49.4	Average to very long
Arm length (GF)	$\dfrac{100 \times 35 +/- 1}{180 +/- 2} = 18.5$ to 20.2	Short to long (?)

Arm length (ET)	$\dfrac{100 \times 35 +/- 1}{184 +/- 2} = 18.2$ to 19.7	Short to average
Forearm length (GF)	$\dfrac{100 \times 27 +/- 1}{180 +/- 2} = 14.2$ to 15.7	Short to average
Forearm length	(ET) $\dfrac{100 \times 27 +/- 1}{184 +/- 2} = 13.9$ to 15.3	Very short to average
Relative hand length (GF)	$\dfrac{100 \times 26 +/- 0.5}{180 +/- 2} = 14.0$ to 14.8	Abnormally long hand
Relative hand length (ET)	$\dfrac{100 \times 26 +/- 0.5}{184 +/- 2} = 13.7$ to 14.5	-
Hand width	$\dfrac{100 \times 12 +/- 0.5}{26 +/- 0.5} = 43.3$ to 49.0	Average to wide
Length of lower limb (GF)	$\dfrac{100 \times 90 +/- 1}{180 +/- 2} = 48.3$ to 51.6	Very to abnormally short
Length of lower limb (ET)	$\dfrac{100 \times 90 +/-1}{184 +/-2} = 49.4$ to 52.7	-
Length of thigh (GF)	$\dfrac{100 \times 51 +/- 1}{180 +/- 2} = 27.4$ to 29.2	Average to very short
Length of thigh (ET)	$\dfrac{100 \times 51 +/- 1}{184 +/- 2} = 26.8$ to 28.5	Short to abnormally short
Length of leg (GF)	$\dfrac{100 \times 37 +/-1}{180 +/- 2} = 19.7$ to 21.2	Short to very short
Length of leg (ET)	$\dfrac{100 \times 37 +/- 1}{184 +/- 2} = 19.3$ to 20.8	-
Length of foot (GF)	$\dfrac{100 \times 26 +/- 1}{180 +/- 2} = 13.7$ to 15.1	Short
Length of foot (ET)	$\dfrac{100 \times 26 +/- 1}{184 +/- 2} = 13.4$ to 14.8	-
Foot width	$\dfrac{100 \times 16 +/-0.5}{26 +/-1} = 57.1$ to 66.0	Abnormally wide
Upper to lower limb	$\dfrac{100 \times 62}{88} = 70.9$	Upper limbs relatively short relative to lower
Brachial	$\dfrac{100 \times 27}{35} = 77.1$	Forearm short relative to upper arm
Crural	$\dfrac{100 \times 37}{51} = 72.5$	Leg short compared to thigh

Original Bibliography for Bernard Heuvelmans'
"The Mystery of the Iceman"

Bader, O.N., 1936. "New paleontological discoveries around Moscow," *Russkyi. Anthrop. Journal*, Moscow, **4**, 471-475 (in Russian).

Bader, O.N., 1950. "Discovery of the skull cap of a Neanderthal old man around Chvalynsk and the question of its age," *Bull. Mosk. ob-va ispitatet. prorody (Otdel geologichetsky)*, Moscow, **18**, (2), 73-81 (in Russian).

Bontch-Osmolovski, G.A., 1941. "The hand of the fossil man from the Kiik-Koba cave," *Paleolit. Kryma*, vol II, Moscow-Leningrad, Izdat. Akad. Nauk USSR (in Russian).

Bontch-Osmolovski, G.A., 1954. "The skeleton of the foot and of the hand of the fossil man of the Kiik-Koba cave," *ibid.* Vol III (in Russian).

Bordes, François, 1961. "Mousterian Cultures in France." *Science*, 134, p 803-810.

Boule, Marcellin, 1912-13. "L'homme fossile de la Chapelle aux Saints," Extracts from *Annales de Paléontologie*, Masson, Paris.

Boule, M. and H.V. Vallois, 1952. *Les Hommes Fossiles*. 4th ed., Masson, Paris.

Bounak, V.V., 1954. "Current status of the problem of the evolution of the foot among the ancestors of Man," *Paleolit. Kryma*, vol. III Moscow-Leningrad, Izdat. Akad. Nauk SSSR (in Russian).

Brace, C. Loring., 1968. "Ridiculed, Rejected, but still our Ancestor: Neanderthal," *Natural History*, New York, May 1968, p 38-45.

Britton, Sydney W., 1955. "Man walks upright because of snow," *Science News*, Washington, April 23.

Broom, Robert, 1934. *Les origines de l'homme*. Payot, Paris.

Burchett, Wilfred C., *La Seconde Resistance, Viet-nam 1965*. [Chapter: "Du Yéti aux Eléphants."] Gallimard, Paris.

Coon, Carleton S., 1962. *The Origin of Races*. Alfred E. Knopf, New York.

Cordier, Charles, 1963. "Deux anthropoides inconnus marchant debout, au Congo ex-Belge," *Genus*, **29**, Nos 1-4, Rome.

Cordier, Charles, 1973. "Animaux inconnus du Congo," *Zoo*, **38**, (4), p. 185-191, Antwerp.

Dappert, Olfert, 1668. *Naukeurige Beschrijvinge der Afrikaensche Geweste van Egypten, Barbaryen, Libyen, Biledulgerid, Negroslant, Guinea, Ehtiopien, Abyssinie.* Wolfgang, Waesberge, Boom en Van Someren, Amsterdam.

Dappert, Olfert, 1686. *Description de l'Afrique.* Wolfgang, Waesberge, Boom en Van Someren, Amsterdam.

Dart, Raymond A., 1957. "The Makapansgat Austalopithecine Osteodontokeratic Culture," *Proc. Third Pan-Afric. Congress on Prehistory held in N. Rhodesia.* London, p. 161-171.

Darwin, Charles, 1871. *The Descent of Man, and Selection in Relation to Sex.* John Murray, London.

DeBeer, Gavin R., 1951. *Embryos and Ancestors*, 2nd ed. Oxford Univ. Press, New York.

DuBrul, E.L. and Charles A. Reed, 1960. "Skeletal Evidence of Speech?" *Amer. Journ. Phys. Anthrop.* **18**, no 2, p. 153-156.

Egorov, N.M., 1933. "Zur Frage über das Alter der sogenannten Podkumok-Menschen," *Anthrop. Anzeiger*, Stuttgart, **10**, p. 223-225.

Faure, Maurice, 1923. "Reconstitution de l'Homo mousteriensis ou neanderthalensis," *Assoc. Franç. Avancement des Science, Congrès de Montpellier, 1922.* pp 439-444.

Frechkop, Serge, 1949. "Le crâne de l'homme en tant the crâne de mammifère," *Bull. Inst. roy. Sci. nat. Belgique*, **26**, no 23.

Frechkop, Serge, 1954a. "Le port de la tête et la forme du crâne chez les Singes," *ibid.* **30**, no. 12.

Frechkop, Serge, 1954b. "Professor Max Westenhöfer on the Problem of Man's Origin," *Eugenics Review*, **46**, no. 1, p. 42-48. London.

Gaudry, Albert, 1878-90. *Les Enchaînements du Monde Animal.* Savy, Paris.

Goudelis, V. and S. Pavilonis, 1952. "The skull of fossil Man," *Priroda*, **6**, Leningrad (in Russian).

Goury, Georges, 1948. *Origine et Evolution de l'Homme.* A. et J. Picard, Paris.

Gremiatski, M.A., 1922. "The Podkumok skull cap and its morphological features," *Russkyi. Anthrop. Journal*, Moscow, **12**, no 1 -2, p 92-100 and 237-239 (in Russian).

Gerasimov, Mikhail, M. 1971. *The Face Finder.* Hutchinson, London.

Haeckel, Ernst, 1866. *Generelle Morphologie der Organismen.* Zweiter Bund, p 300, Berlin.

Heuvelmans, Bernard, 1954a. "D'après les travaux les plus récents, ce n'est pas l'Homme qui descend du Singe, mais le Singe qui descendrait de l'Homme," *Science et Avenir,* Paris, no 84, p 58-61, 96.

Heuvelmans, Bernard, 1954b. "L'Homme doit-il être considéré comme le moins spécialisé des Mammifères?" *ibid.,* no 85, p 132-136, 139.

Heuvelmans, Bernard, 1969. "Note préliminaire sur un spécimen conservé dans la glace, d'une forme encore inconnue d'Hominidé vivant: *Homo pongoïdes* (sp. seu subsp. nov.)," *Bull. Inst. roy. Sci. nat. Belgique,* **45**, no. 4.

Hill, W.C. Osman, 1945. "Nittaewo, an Unsolved Problem in Ceylan," *Loris,* Colombo, 4, p. 241-262.

Hooton, Ernest A., 1946. *Up from the Ape.* Macmillan, New York.

Howell, F. Clark, 1965. *Early Man.* Time Inc., New York.

Howells, William, 1944. *Mankind So Far.* Doubleday, Doran & Co., New York.

Hulse, Frederick. S., 1963. *The Human Species.* Random House, New York.

Koenigswald, G.H.R. von, Editor, 1958. *Hundert Jahre Neanderthaler-Neanderthal Centenary 1856-1956.* Böhlau, Köln und Graz.

Koffman, Marie-Jeanne, 1967. "Preliminary appraisal of the problem of relic hominoids in the Caucasus," Akad. Sci. USSR, geographical division, Thesis and conference on a problem of medical geography in the northern Caucasus, Leningrad, 1967 (in Russian).

Koffman, Marie-Jeanne, 1968. "Footprints remain..." *Nauka i Religia,* no 4 (in Russian).

Kollman, J., 1906. "Der Schädel von Kleimkems und die Neanderthal-Spy Gruppe," *Archiv für Anthropologie,* Braunschweig, **5**, Heft 3, p. 208.

Kortlandt, Adriaan, 1962. "Chimpanzees in the Wild," *Scientific American,* 206, p 128-138.

Kortlandt, Adriaan, and M.Kooij, 1963. "Protohominid Behavior in Primates," *Symposia Zool. Soc. London,* **10**, p 61-88, May 1962.

Kurten, Björn, 1971. *Inte från aporna.* Albert Bonniers Förlag, Stockholm.

Kurten, Björn, 1972. *Not From the Apes*. Victor Gollancz, London.

Kurth, G., 1958. "Betrachtungen zu Rekonstructionsversuchen," in Koengiswald, 1958, p. 217-230, Panel LVI.

Leakey, L. S. B., 1971. in "Australopithecus, a long-armed short-legged, knuckle-walker," *Science News*, Washington, **100**, no 2, p. 357. Nov. 27.

Le Double, A.-F. et F. Houssay, 1912. *Les Velus*. Vigot Frères, Paris.

LeGros Clark, W.C., 1958. *History of the Primates,* 6th ed. British Museum Nat. History, London.

LeGros Clark, W.C., 1964. *The Fossil Evidence for Human Evolution,* 2nd ed. Univ. Chicago Press.

Lieberman, Philip and Edmund S. Crelin, 1971. "On the Speech of Neanderthal Man," *Linguistic Enquiry*, **2**, no 2.

Linnaeus, Carl von, 1758. *Systema Naturae*, 10th ed. Holmae, Salvius.

Lorenz, Konrad, 1959. "Psychologie und Stammesgeschichte," in Heberer, G. *Evolution der Organismen*. Fisher, Stuttgart.

Lorenz, Konrad, 1963. *Das Sogenannte Böse*. Borotha-Schoeler.

Loth, Edward, 1931. *L'Anthropologie des Parties Molles*. Masson, Paris.

Loth, Edward, 1936. "Considérations sur les reconstructions des muscles de la race du Nanderthal," *C.R. de l'Assoc. des Anatom*. Mila, 3-8 Sept. 1936.

Loth, Edward, 1938. "Beitrage zur Kentniss des Weichteilanatomie des Neanderthalers," *Zeitschrift fr Rassenkunde*, Stuttgart, **7**, p 13-35.

Lounine, B. V., 1937. "On the question of the actual age of 'Podkumuk Man' in the light of archeological data," *Sovietskaya Archeologia*, Moscow, **4**, p 63-90.

McGregor, J.H., 1926. "Restoring Neanderthal Man," *Natural History*, New York, **26**, no. 3, p. 228-293.

Malson, Louis, 1964. *Les Enfants Sauvages*. Union Gnrales d'Editions, Paris.

Mauduit, Jacques A., 1954. *Quarante mille ans d'Art Moderne*. Plon, Paris.

Mayr, Ernst, 1963. "The taxonomic Evaluation of Fossil Hominids," in Washburn, S.L. *Classification and Human Evolution.* Aldine, Chicago.

Morant, G.M., 1927. "Studies of Paleolithic Man: II A biometric study of Neanderthaloid skulls and their relationships to modern racial types," *Ann. Eugenics,* London, **2**, p 310-380.

Morris, Desmond, 1967. *The Naked Ape.* Jonathan Cape, London.

Napier, John R., 1962. "The Evolution of the Human Hand," *Scientific American,* **207,** no 6.

Nizami al'Aroudi al Samarqandi, 1921. *Revised translation of the Chahar Maqala ('Four Discourses') of Nizami-i- Arudi of Samarqand.* Luzde, London.

Osborn, Henry F., 1927. "The Origin and Antiquity of Man: a Correction," *Science,* Washington, **65**, no 1694 p 597.

Osborn, Henry F., 1929. "Is the Ape-Man a Myth*?"* *Human Biology,* Detroit, **1,** no 1, p 4-9.

Pales, Lon, 1960. "Les empreintes de pieds humains de la Tana della Basura," *Revue d'Etudes Ligures,* Bordighera, **26**, nos 1-4.

Patte, Etienne, 1955. *Les Nanderthaliens.* Masson, Paris.

Pavlov, A.P., 1925. "Fossil man from the Mammoth Era in eastern Russia and fossil men of western Europe," *Russkyi. Anthrop. Journal,* Moscow, **14**, no 1-2, p. 5-36.

Piveteau, Jean, 1957. *Traité de Paléontologie.* (vol VII: *Paléontologie Humaine, Les Primates et les Hommes.)* Masson, Paris.

Porshnev, Boris F., 1966. "L'Origine de l'Homme et les Hominoides velus," *Genus,* Rome, ser. IIa, **2**, no. 3.

Porshnev, Boris F., 1968. "The Struggle for the Troglodytes," *Prostor,* Alma-Ata, nos 4-7 (in Russian).

Porshnev, Boris F., 1969a. "The Problem of the relic Paleanthropes," *Sovetskaya Ethnographia,* Moscow, no 2, p 115-130 (in Russian).

Porshnev, Boris F., 1969b. "The Troglodytes and the Hominids in the systematics of the higher Primates,"*Doklady Akad. Nauk, SSSR,* Moscow, **188**, no 1 (in Russian).

Porshnev, Boris F., 1969c. "A Paleanthrope?" *Technika Molodezhi,* Moscow, **11**, p 34-36 (in Russian).

Porshnev, Boris F., 1971. "The second signaling system as a criterion for differentiating between Troglodytes and Hominids," *Doklady Akad. Nauk, SSSR*, Moscow, **190**, no 1.

Ranke, Johann, 1897. " Die individuellen Variationen in Schädelbau des Menschen," *Korr.- Bl. d. Deutschen Geselschaft f. Archaeologie, Ethnologie u. Urgeschichte*, München, **28**, p 134-146.

Rengarten, V.P., 1922. "On the age of the sediments which contained the remains of Podkoumuk Man," *Russkyi. Anthrop. Journal*, Moscow, **12**, p 193-195.

Richardson, L.F., 1960. *Statistics of Deadly Quarrels*. Stevens and Sons, London.

Sanderson, Ivan T., 1950. "Abominable Snowman," *True Magazine*, New York, May.

Sanderson, Ivan T., 1961. *Abominable Snowmen: Legend Come to Life*. Chilton Co., Philadelphia and New York.

Sanderson, Ivan T., 1967. "The Wudewsa or Hairy Primitives of Ancient Europe," *Genus*, Rome, **23**, no 1-2.

Servant, Michel, P. Ergenzinger and Y. Coppens, 1969. "Datations absolues sur un delta lacustre quaternaire au Sud du Tibesti (Angamma)," *C. R. S. Soc. Gol. Fr.*, Paris.

Severtsov, A.N. 1939. *Evolution's regular morphological development*. Izdat. Akad. Nauk SSSR, Moscow-Leningrad (in Russian).

Schultz, A.H., 1931. "The density of hair in Primates," *Human Biology*, Detroit, **3**, p 303-321.

Schultz, A.H., 1955. "Das Bild ausgestorbener Menschen," *Umschau*, Frankfurt, Leipzig, **5,** p 143-145.

Stolywho, Kasimierz, 1908. "*Homo primigenius* appartient-il une espèce distincte de l'Homo sapiens?" *L'Anthropologie*, Paris, 19, p 191-216.

Symington 1960: in DuBrul and Reed.

Tyson, Edward, 1699. *Orang-utang, sive Homo sylvestris: or the anatomy of a pygmie compared with that of a monkey, an ape and a man*. Thomas Bonnet, London.

Ullrich, H., 1958. "Neandertaler funde aus der Sowjet Union," (in Koenigswald, 1958, p 72-102).

Vallois, Henri V., 1949. "L'origine de l'Homo sapiens," *C.R. Acad. Sci. Paris.* **228**.

Vallois, Henri V., 1954. "Neandertals and Praesapiens," *J. Roy. Anthrop. Inst.* London, **84.**

Vallois, Henri V., 1969. "Le temporal néandertalien H27 de La Quina: Etude anthropologique," *L'Anthropologie*, Paris, **73**, no 5-6, p 365-400; no 7-8, p 525-544.

Vollmer, 1828. *Natur- und Sittengemälde der Tropenländer* (p 224-226, plate VI) G. Michaelis, München.

Weinert, H. 1944. *L'Homme préhistorique.* Payot, Paris.

Westenhfer, Max, 1942. *Der Eigenweg des Menschen.* Die Medizinische Welt, Berlin.

Westenhfer, Max, 1948. *Die Grunlagen meiner Theorie von Eigenweg des Menschen.* C. Winter, Heidelberg.

Westenhfer, Max, 1953. *Le Problème de la Genèse de l'Homme.* Office de la Publicité, E. Sobeli, Bruxelles.

Zeuner, F.E., 1958. "The replacement of Neanderthal Man by *Homo sapiens*," in Koenigswald, 1958, p 312-316.

ON THE MATTER OF THE ICEMAN (in chronological order)

Note: The only publications listed here are those that provide some new information and which, altogether, contribute to the understanding of the sequence of events.

Lucas, Jim, G. 1966. "Hunger Keeps Marines Sharp," *World Journal Tribune*, New York, Nov. 1, 1966.

Heuvelmans, Bernard, 1969. "Note préliminaire sur un spécimen conservé dans la glace, d'une forme encore inconnue d'Homindé vivant: *Homo pongoïdes* (sp. seu subsp. nov.)," *Bull. Inst. roy. Sci. nat. Belgique,* **45**, no. 4.

Burnet, Albert, 1969. "L'Homme de Néanderthal serait toujours vivant," (First reaction in the international press] *Le Soir*, Bruxelles, March 11.

Anon. 1969. "Cool reaction for 'Neanderthal Man,'" *New Scientist*, March 20, 1969.

Linklater, Magnus, 1969. "Is it a fake?... is it an ape?...or is it...Neanderthal Man?" *Sunday Times*, London, March 27.

Casey, Phil, 1969a. "New race of Man? Carnival Exhibit Interests Smithsonian," *Washington Post*, March 27.

Schaden, Herman, 1969. "The Rambler...takes Up a Mystery," *Evening Star*, Washington, D.C. March 27.

Jones, Ellsworth, and I. MacBeath, 1969. "Ape-man examination refused," *Sunday Times*, London, March 30.

Casey, Phil, 1969b. "The Smithsonian Courts a Thing," *New York Post*, April 11.

Anon. "Call to FBI on 'freak' in ice block," *Times*, London, April 12.

Casey, Phil, 1969c. "Ice Man Origin Solid Mystery to Smithsonian," *Los Angeles Times*, April 17.

Stettler, Hans, 1969. "Der Neanderthaler lebt!" *Blick*, Zurich, April 1969.

Sanderson, Ivan T., 1969. "Editorial,*" Pursuit,* Columbia, N.J. **2**, no2, April.

Sanderson, Ivan T., 1969. "The Missing Link," *Argosy*, New York, May.

Anon. 1969. "Expert hints Ice Man was Custom-Built," *New York Post*, May 10.

Anon. 1969. "The Iceman Melteth." *Scientific Research*, p. 17, May 26.

Heuvelmans, Bernard, 1969. "Neanderthal Man," *Personality*, Johannesburg, June 5.

Sanderson, Ivan T., 1969. "Ends 'Bozo', we think," *Pursuit,* Columbia, N.J., July.

McCrystal, Cal, 1969. "The Iceman's magical mystery tour," *Sunday Times*, London, Sept. 28.

Sanderson, Ivan T., 1969. "Preliminary Description of the External Morphology of What Appeared to be the Fresh Corpse of a Hitherto Unknown Form of Living Hominid," *Genus*, Rome, **25**, no 1-4, p 249-289.

Porshnev, Boris F., 1969. "A Paleanthrope?" *Technika Molodezhi*, Moscow, p 34-37, Nov.

Anon. 1969. "O abominvel Homem do glo," *Manchete*, Rio de Janeiro, p 38-41, Dec 27.

Heuvelmans, Bernard, 1970. "Bestia, Hombre or Estabon Perdido?" *Excelsior*, Mexico, Jan 15, 17, 18.

Hansen, Frank. D., 1970. "I Killed the Ape-Man Creature of Whiteface," *Saga*, New York, July.

Herrier, Robert, 1970. "J'ai vu l'Homme-singe, affirme le professeur Heuvelmans," *Noir et Blanc*, Paris, Aug. 24-30.

Vogel, Eric, 1970. "Rocambolesque, incroyable: un 'homme sauvage,' congelé," *Tribune de Genève*, Aug 22-23.

Vogel, Eric, 1970. "Des hommes de la période glaciaire vivent-ils encore de nos jours?" *ibid*, Oct. 10.

Hall, Tom, 1971. "Tracking the Minnesota monster," *Chicago Tribune Sunday Magazine*, sect. 7, p 18-36, Oct. 24.

Napier, John, 1972. "Bigfoot: The Yeti and Sasquatch in Myth and Reality," Chap. 4, in *Tales from the Minnesota Woods*, Jonathan Cape, London.

McCoy, Alfred W. 1972 *"The Politics of Heroin in Southeast Asia*. Harper and Row, New York.

Lamour, Catherine and M.R. Lambert, 1972. *Les Grandes Maneuvres de l'Opium*. Editions du Seuil, Paris.

Anon. 1972a. "Suspect Held in Ring Hiding Dope in Bodies," *Philadelphia Enquirer*, Dec. 23.

Anon. 1972b. "Phony ID is Tied to Drug Ring That Used Bodies of War Dead," *ibid*. Dec. 24.

Anon. 1973a. "U.S. Indicts Phony GI In Heroin Case," *Evening and Sunday Bulletin*, Philadelphia, Jan. 3.

Anon, 1973b. "Probers Eye Civilians in GI Body Drug Plot," *Philadelphia Enquirer*, Jan. 4.

Buckley, Thomas, 1973. "How the C.I.A. Got Hooked on Heroin," *Penthouse Magazine*, June.

Anon. 1973b. "Anti-Drug Chief Quits Post, Hits White House Interference," *International Herald Tribune*, June 30 -July 1.

CPSIA information can be obtained
at www.ICGtesting.com
Printed in the USA
BVHW011626100720
583375BV00012B/497

9 781938 398810